HLSL Development Cookbook

Implement stunning 3D rendering techniques using the
power of HLSL and DirectX 11

Doron Feinstein

PUBLISHING

BIRMINGHAM - MUMBAI

HLSL Development Cookbook

First published: June 2013

Production Reference: 1060613

Published by Packt Publishing Ltd.
Livery Place
35 Livery Street
Birmingham B3 2PB, UK.

ISBN 978-1-84969-420-9

www.packtpub.com

Cover Image by Artie Ng (artherng@yahoo.com.au)

Credits

Author
Doron Feinstein

Reviewers
Brecht Kets

Pope Kim

Acquisition Editor
Erol Staveley

Lead Technical Editors
Arun Nadar

Hardik Saiyya

Technical Editor
Kaustubh S. Mayekar

Project Coordinator
Arshad Sopariwala

Proofreader
Mario Cecere

Indexer
Tejal R. Soni

Graphics
Ronak Dhruv

Abhinash Sahu

Valentina D'silva

Production Coordinator
Shantanu Zagade

Cover Work
Shantanu Zagade

About the Author

Doron Feinstein has been working as a graphics programmer over the past decade in various industries. Since he graduated his first degree in software engineering, he began his career working on various 3D simulation applications for military and medical use. Looking to fulfill his childhood dream, he moved to Scotland for his first job in the game industry at Realtime Worlds. Currently working at Rockstar Games as a Senior Graphics Programmer, he gets to work on the company's latest titles for Xbox 360, PS3, and PC.

I would like to thank my wife, who convinced me to write this book, for all her support.

About the Reviewers

Brecht Kets is a Senior Lecturer at Howest University in Belgium, where he teaches game development in one of the leading international game development study programs, Digital Arts, and Entertainment (www.digitalartsandentertainment.com). He's been actively involved in game development for several years, and has been writing about XNA since the launch in December 2006. He hosts the www.3dgameprogramming.net website and has received the Microsoft Most Valuable Professional award in the category DirectX/XNA six times in a row for his contributions in the community.

He has also co-authored the book *Building your First Mobile Game using XNA 4.0*, *Packt Publishing* and the video series *XNA 3D Game Development By Example*, *Packt Publishing*.

Pope Kim is a seasoned rendering programmer with over 10 years of experience in the gaming industry. While working with top game studios in the world, he has shipped over a dozen games on many platforms, including Xbox 360, PS3, PC, Wii, PS2, and PSP.

He has degrees in Law and Computer Science, and is an occasional presenter at computer graphics or game-related conferences, such as Siggraph and Korea Game Conference.

He is also a part-time educator. He served his 3 years at the Art Institute of Vancouver as an HLSL programming instructor and currently holds a professor position at Sogang University Game Education Center.

In 2012, he authored an introductory HLSL programming book, which instantly became a best-seller in Korea. It is currently being translated back to English and is partly available on his blog.

You can follow Pope at http://www.popekim.com or on Twitter at @BlindRenderer.

www.PacktPub.com

Support files, eBooks, discount offers and more

You might want to visit www.PacktPub.com for support files and downloads related to your book.

Did you know that Packt offers eBook versions of every book published, with PDF and ePub files available? You can upgrade to the eBook version at www.PacktPub.com and as a print book customer, you are entitled to a discount on the eBook copy. Get in touch with us at service@packtpub.com for more details.

At www.PacktPub.com, you can also read a collection of free technical articles, sign up for a range of free newsletters and receive exclusive discounts and offers on Packt books and eBooks.

http://PacktLib.PacktPub.com

Do you need instant solutions to your IT questions? PacktLib is Packt's online digital book library. Here, you can access, read and search across Packt's entire library of books.

Why Subscribe?

- ▶ Fully searchable across every book published by Packt
- ▶ Copy and paste, print and bookmark content
- ▶ On demand and accessible via web browser

Free Access for Packt account holders

If you have an account with Packt at www.PacktPub.com, you can use this to access PacktLib today and view nine entirely free books. Simply use your login credentials for immediate access.

Table of Contents

Preface

DirectX 11 has been around for a couple of years now but never received much attention from 3D developers up until now. With PC regaining its popularity in the gaming community and the third generation of Xbox gaming console just around the corner, the transition to DirectX 11 is just a matter of time.

In this book, we will cover common and new 3D techniques implemented using the features offered by DirectX 11. From basic lighting to advanced screen space effects, each recipe will introduce you to one or more new DirectX 11 features such as Compute Shaders, Unordered Access Views, and Tessellation.

The *HLSL Development Cookbook* will provide you with a series of essential recipes to help you make the most out of the different rendering techniques used within games and simulations using the DirectX 11 API.

What this book covers

Chapter 1, Forward Lighting, will guide you through the light equation for the most commonly used light sources implemented in the forward lighting technique.

Chapter 2, Deferred Shading, will teach you to optimize the lighting calculations introduced in the previous chapter by separating the light calculations from the scene complexity.

Chapter 3, Shadow Mapping, will help you to improve the realism of the lighting calculations covered in the previous chapters by adding shadow support.

Chapter 4, Postprocessing, will guide you to enhance the quality of the lighting results with different 2D filtering techniques.

Chapter 5, Screen Space Effects, will help you to extend the lighting calculations from *Chapter 2, Deferred Lighting*, by implementing additional lighting effects as screen space calculations.

Chapter 6, Environment Effects, will help you to add some final touches to your rendered scene by changing the weather and adding some interactivity.

What you need for this book

Running the samples provided with this book requires a computer with a DirectX 11-enabled graphics card running Windows Vista or newer operating system. Compiling the code will require Microsoft's Visual Studio 2010 or newer with DirectX SDK (June 2010).

Who this book is for

If you have some basic Direct3D knowledge and want to give your work some additional visual impact by utilizing the advanced rendering techniques, then this book is for you. It is also ideal for those seeking to make the transition from DirectX 9 to DirectX 11 and those who want to implement powerful shaders with the HLSL.

Conventions

In this book, you will find a number of styles of text that distinguish between different kinds of information. Here are some examples of these styles, and an explanation of their meaning.

Code words in text are shown as follows: "Once the constant buffer is updated, bind it to the pixel shader using the context function `PSSetConstantBuffers`."

A block of code is set as follows:

```
cbuffer HemiConstants : register( b0 )
{
    float3 AmbientDown    : packoffset( c0 );
    float3 AmbientRange   : packoffset( c1 );
}
```

New terms and **important words** are shown in bold. Words that you see on the screen, in menus or dialog boxes for example, appear in the text like this: "Where α is the angle between **N** and **L**".

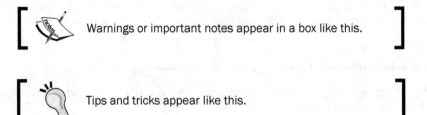

Warnings or important notes appear in a box like this.

Tips and tricks appear like this.

Reader feedback

Feedback from our readers is always welcome. Let us know what you think about this book—what you liked or may have disliked. Reader feedback is important for us to develop titles that you really get the most out of.

To send us general feedback, simply send an e-mail to feedback@packtpub.com, and mention the book title via the subject of your message.

If there is a topic that you have expertise in and you are interested in either writing or contributing to a book, see our author guide on www.packtpub.com/authors.

Customer support

Now that you are the proud owner of a Packt book, we have a number of things to help you to get the most from your purchase.

Downloading the example code

You can download the example code files for all Packt books you have purchased from your account at http://www.packtpub.com. If you purchased this book elsewhere, you can visit http://www.packtpub.com/support and register to have the files e-mailed directly to you.

Downloading the color images of this book

We also provide you a PDF file that has color images of the screenshots/diagrams used in this book. The color images will help you better understand the changes in the output.

You can download this file from http://www.packtpub.com/sites/default/files/downloads/4209OT_ColoredImages.pdf.

Errata

Although we have taken every care to ensure the accuracy of our content, mistakes do happen. If you find a mistake in one of our books—maybe a mistake in the text or the code—we would be grateful if you would report this to us. By doing so, you can save other readers from frustration and help us improve subsequent versions of this book. If you find any errata, please report them by visiting http://www.packtpub.com/submit-errata, selecting your book, clicking on the **errata submission form** link, and entering the details of your errata. Once your errata are verified, your submission will be accepted and the errata will be uploaded on our website, or added to any list of existing errata, under the Errata section of that title. Any existing errata can be viewed by selecting your title from http://www.packtpub.com/support.

Piracy

Piracy of copyright material on the Internet is an ongoing problem across all media. At Packt, we take the protection of our copyright and licenses very seriously. If you come across any illegal copies of our works, in any form, on the Internet, please provide us with the location address or website name immediately so that we can pursue a remedy.

Please contact us at copyright@packtpub.com with a link to the suspected pirated material.

We appreciate your help in protecting our authors, and our ability to bring you valuable content.

Questions

You can contact us at questions@packtpub.com if you are having a problem with any aspect of the book, and we will do our best to address it.

1
Forward Lighting

In this chapter we will cover:

- Hemispheric ambient light
- Directional light
- Point light
- Spot light
- Capsule light
- Projected texture – point light
- Projected texture – spot light
- Multiple lights in a single pass

Introduction

Forward lighting is a very common method to calculate the interaction between the various light sources and the other elements in the scene, such as meshes and particle systems. Forward lighting method has been around from the fixed pipeline days (when programmable shaders were just an insightful dream) till today, where it gets implemented using programmable shaders.

From a high-level view, this method works by drawing every mesh once for each light source in the scene. Each one of these draw calls adds the color contribution of the light to the final lit image shown on the screen. Performance wise, this is very expensive—for a scene with N lights and M meshes, we would need N times M draw calls. The performance can be improved in various ways. The following list contains the top four commonly used optimizations:

- Warming the depth buffer with all the fully opaque meshes (that way, we don't waste resources on rendering pixels that get overwritten by other pixels closer to the camera).

▶ Skip light sources and scene elements that are not visible to the camera used for rendering the scene.

▶ Do bounding tests to figure which light affects which mesh. Based on the results, skip light/mesh draw calls if they don't intersect.

▶ Combine multiple light sources that affect the same mesh together in a single draw call. This approach reduces the amount of draw calls as well as the overhead of preparing the mesh information for lighting.

Rendering the scene depths, as mentioned in the first method, is very easy to implement and only requires shaders that output depth values. The second and third methods are implemented on the CPU, so they won't be covered in this book. The fourth method is going to be explained at the end of this chapter. Since each one of these methods is independent from the others, it is recommended to use all of them together and gain the combined performance benefit.

Although this method lost its popularity in recent years to deferred lighting/shading solutions (which will be covered in the next chapter) and tiled lighting due to their performance improvement, it's still important to know how forward lighting works for the following reasons:

▶ Forward lighting is perfect for lighting scene elements that are not fully opaque. In fact, both deferred methods only handle opaque elements. This means that forward lighting is still needed for scenes containing translucent elements.

▶ Forward lighting can perform well when used for low-quality rendering tasks, such as low-resolution reflection maps.

▶ Forward lighting is the easiest way to light a scene, which makes it very useful for prototyping and in cases where real-time performance is not important.

All the following recipes are going to cover the HLSL side of the rendering. This means that you, the reader, will need to know how to do the following things:

▶ Compile and load the shaders

▶ Prepare a system that will load and manage the scene

▶ Prepare a framework that supports Direct3D draw calls with shaders that will render the scene

All vertex buffers used with this technique must contain both positions and normals. In order to achieve smooth results, use smooth vertex normals (face normals should be avoided).

In addition, the pixel shader has to come up with a per-pixel color value for the rendered meshes. The color value may be a constant per mesh color or can be sampled from a texture.

Hemispheric ambient light

Ambient light is the easiest light model to implement and yet it is very important to the overall look of your scene. For the most part, ambient light refers to any light in the scene that cannot be directly tied to a specific light source. This definition is flexible and its implementation will be shown soon.

In the past, a single constant color value was used for every mesh in the scene that provides a very flat result. As programmable shaders became more available, programmers switched from constant color to other solutions that take the mesh normal into account and avoid the flat look. Hemispheric lighting is a very common method to implement ambient lighting that takes normal values into account and does not require a lot of computations. The following screenshot shows the same mesh rendered with a constant ambient color (left-hand side) and with hemispheric lighting (right-hand side):

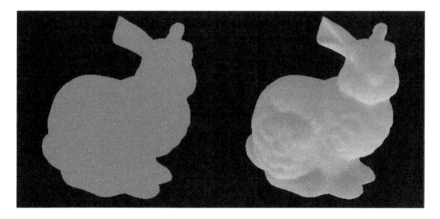

As you can see, constant ambient light hides all the mesh detail, while hemispheric light provides a much more detailed result.

Getting ready

Hemispheric ambient light requires two colors that represent the light coming from above and below each mesh being rendered. We will be using a constant buffer to pass those colors to the pixel shader. Use the following values to fill a D3D11_BUFFER_DESC object:

Constant Buffer Descriptor Parameter	Value
Usage	D3D11_USAGE_DYNAMIC
BindFlags	D3D11_BIND_CONSTANT_BUFFER
CPUAccessFlags	D3D11_CPU_ACCESS_WRITE
ByteWidth	8

The reset of the descriptor fields should be set to 0.

Creating the actual buffer, which is stored as a pointer to a `ID3D11Buffer` object, call the D3D device function `CreateBuffer` with the buffer-descriptor pointer as the first parameter, `NULL` as the second parameter, and a pointer to your `ID3D11Buffer` pointer as the last parameter.

How to do it...

All lighting calculations are going to be performed in the pixel shader. This book assumes that you have the basic knowledge to set up and issue the draw call for each mesh in the scene. The minimum calculation a vertex shader has to perform for each mesh is to transform the position into projected space and the normal into world space.

 If you are not familiar with the various spaces used in 3D graphics, you can find all the information you will need on Microsoft's MSDN at http://msdn.microsoft.com/en-us/library/windows/desktop/bb206269%28v=vs.85%29.aspx.

As a reference, the following vertex shader code can be used to handle those calculations:

```
cbuffer cbMeshTrans : register( b0 )
{
  float4x4  WorldViewProj  : packoffset( c0 );
  float4x4  World     : packoffset( c4 );
}

struct VS_INPUT
{
  float4 Pos   : POSITION;
  float3 Norm  : NORMAL;
  float2 UV  : TEXCOORD0;
};

struct VS_OUTPUT
{
  float4 Pos   : SV_POSITION;
  float2 UV  : TEXCOORD0;
  float3 Norm  : TEXCOORD1;
};

VS_OUTPUT RenderSceneVS(VS_INPUT IN)
{
```

```
VS_OUTPUT Output;

// Transform position from object to projection space
Output.Pos = mul(IN.Pos, WorldViewProj);

// Copy the texture coordinate through
Output.UV = input.TextureUV;

// Transform normal from object to world space
Output.Norm = mul(IN.Norm, (float3x3)World);

return Output;
}
```

Downloading the example code

You can download the example code files for all Packt books you have purchased from your account at http://www.packtpub.com. If you purchased this book elsewhere, you can visit http://www.packtpub.com/support and register to have the files e-mailed directly to you.

Again, this code is for reference, so feel free to change it in any way that suits your needs.

In the pixel shader, we will use the following deceleration to access the values stored in the constant buffer:

```
cbuffer HemiConstants : register( b0 )
{
   float3 AmbientDown    : packoffset( c0 );
   float3 AmbientRange   : packoffset( c1 );
}
```

See the *How it works...* section for full details for choosing the values for these two constants.

Unless you choose to keep the two constant buffer values constant, you need to update the constant buffer with the values before rendering the scene. To update the constant buffer, use the context functions, Map and Unmap. Once the constant buffer is updated, bind it to the pixel shader using the context function PSSetConstantBuffers.

Our pixel shader will be using the following helper function to calculate the ambient value of a pixel with a given normal:

```
float3 CalcAmbient(float3 normal, float3 color)
{
   // Convert from [-1, 1] to [0, 1]
   float up = normal.y * 0.5 + 0.5;
```

```
    // Calculate the ambient value
    float3 Ambient = AmbientDown + up * AmbientUp;

    // Apply the ambient value to the color
    return Ambient * color;
}
```

This function assumes the normal y component is the up/down axis. If your coordinate system uses a different component as the vertical axis, change the code accordingly.

Similar to the vertex shader, the code for the pixel shader entry point depends on your specific mesh and requirements. As an example, the following code prepares the inputs and calls the helper function:

```
    // Normalize the input normal
    float3 normal = normalize(IN.norm);

    // Convert the color to linear space
    color = float4(color.rgb * color.rgb, color.a);

    // Call the helper function and return the value
    return CalcAmbient(normal, color);
```

How it works...

In order to understand how ambient light works, it is important to understand the difference between how light works in real life and the way it works in computer graphics. In real life, light gets emitted from different sources such as light bulbs and the Sun. Some of the rays travel straight from the source to our eyes, but most of them will hit surfaces and get reflected from them in a different direction and with a slightly different wave length depending on the surface material and color. We call each time the light gets reflected from a surface to a bounce. Since each light bounce changes the wave length, after a number of bounces the wave length is no longer visible; so what our eyes see is usually the light that came straight from the source plus the light that bounced for a very small amount of times. The following screenshot demonstrates a situation where a light source emits three rays, one that goes directly to the eye, one that bounces once before it reaches the eye, and one that bounces twice before it reaches the eye:

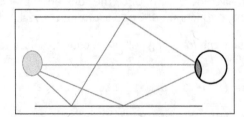

In computer graphics, light calculation is limited to light that actually reaches the viewer, which is usually referred to as the camera. Calculating the camera's incoming light is normally simplified to the first bounce mostly due to performance restrictions. Ambient light is a term that usually describes any light rays reaching the camera that bounced from a surface more than once. In the old days, when GPUs where not programmable, ambient light was represented by a fixed color for the entire scene.

Graphics Processing Unit (**GPU**) is the electronic part in charge of graphics calculations. When a GPU is not programmable, we say that it uses a fixed pipeline. On the other hand, when a GPU is programmable, we say that it uses a programmable pipeline. DirectX 11-enabled cards are all programmable, so you are not likely to work with a fixed pipeline.

As the first screenshot in this recipe introduction showed, using a fixed color provides a flat and artificial look. As programmable GPUs became commonly available, developers finally had the flexibility to implement better the ambient light models that provide a more natural look. Although the hemispheric ambient light model is not a perfect representation of light that bounced more than once, it gained its popularity due to its simplicity and quality.

Hemispheric ambient light splits all light rays that affect the mesh getting rendered to those that arrived from above and below the mesh. Each one of these two directions is assigned a different color and intensity. To calculate the ambient light value of a given pixel, we use the normal vertical direction to linearly blend the two colors. As an example, in an outdoor scene with blue sky and grassy ground, the ambient light interpolates across the hemisphere, as shown in the following image:

Picking a pair of colors that properly represents the mesh surrounding for the upper and lower hemisphere is probably the most important step in this recipe. Though you can write code that picks the color pairs based on the scene and the camera position, in most games the values are handpicked by artists.

> Note that even though the color pairs are constant for all the pixels affected by the draw call, they don't have to be constant overtime or for all the meshes in the scene. In fact, changing the color values based on the time of day or room properties is a very common practice.

One thing to keep in mind when picking the colors is the space they are in. When an artist manually picks a color value, he usually comes up with color values in what is known as gamma space. Light calculations on the other hand should always be performed in linear space. Any color in gamma space can be converted to linear space by raising it to the power of 2.2, but a faster and common approximation is to square the color (raising it to the power of 2). As you can see in the pixel shader entry point, we converted the pixel color to linear space before passing it to the ambient calculation.

> If you are not familiar with the gamma and linear color space, you should read about gamma correction to understand why it is so important to calculate lighting in linear space in the following link: `http://www.slideshare.net/naughty_dog/lighting-shading-by-john-hable`.

Once you picked the two values and converted them to linear space, you will need to store the colors in the constant buffer as the down color and the range to the upper color. In order to understand this step, we should look at the way the ambient color is calculated inside the helper function. Consider the following linear interpolation equation:

*DownColor * (1-a) + UpColor * a = DownColor + a * (UpColor - DownColor)*

The equation on the left-hand side blends the two colors based on the value, while the equation on the right-hand side does the exact same thing only with the down color and the range between the two colors. The GPU can handle the calculation on the right-hand side with a single instruction (this instruction is called **madd**), which makes it faster than the equation on the left-hand side. Since we use the equation on the left-hand side, you will need to store the upper, minus-lower color into the second constant buffer parameter.

Directional light

Directional light is mainly used for simulating light coming from very large and far light sources, such as the sun and the moon. Because the light source is very large and far, we can assume that all light rays are parallel to each other, which makes the calculations relatively simple.

The following screenshot shows the same model we used to demonstrate ambient light under directional light:

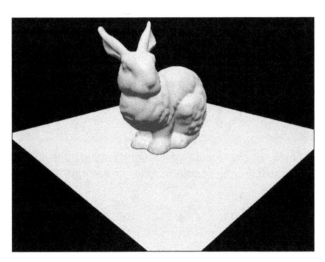

Getting ready

When rendering an outdoor scene that uses directional light to represent the sun or the moon, it is very common to combine the directional light calculation with the ambient light calculation. However, you may still want to support ambient light with no directional light for indoor rendering. For this reason, we will allocate a separate constant buffer for the values used when calculating the directional light. Use the following values in the constant buffer descriptor:

Constant Buffer Descriptor Parameter	Value
Usage	D3D11_USAGE_DYNAMIC
BindFlags	D3D11_BIND_CONSTANT_BUFFER
CPUAccessFlags	D3D11_CPU_ACCESS_WRITE
ByteWidth	8

The reset of the descriptor fields should be set to 0.

The three light values are needed for calculating the directional light: direction, intensity, and color. When rendering a scene with a fixed time of day, those values can be picked in advance by an artist. The only thing to keep in mind is that when this light source represents the Sun/Moon, the sky has to match the selected values (for example, low angle for the Sun means that the sky should show sunset/sunrise).

When time of day is dynamic, you will need multiple values for the different parts of the day/night cycle. An easy way to accomplish that is by picking values for a group of specific times in the day/night cycle (for instance, a value for every 3 hours in the cycle) and interpolate between those values based on the actual position in the cycle. Again, those values have to match the sky rendering.

To apply the light values on a given scene element, a few specific values are needed for the light calculation. Those scene element values will be referred to as the material. The material usually holds per-pixel values such as normal, diffuse color, and specular values. The material values can originate from texture samples or from global values.

How to do it...

Similar to the ambient light, all directional, light-related calculations are handled in the pixel shader. We will be using the following constant buffer declaration in the shader for the new constant buffer:

```
cbuffer DirLightConstants : register( b0 )
{
  float3 DirToLight    : packoffset( c0 );
  float3 DirLightColor : packoffset( c1 );
}
```

Although this may be counterintuitive, the direction used for directional light calculations is actually the inversed direction (direction to the light). To calculate that value, just negate the light direction. The inverted direction is stored in the first shader constant `DirToLight`.

The light intensity value is important when rendering to a **high-dynamic range** (**HDR**) target. HDR is a technique that calculates light values in a range wider than 0 to 1 (for more detail, check the HDR rendering recipe in *Chapter 4, Postprocessing*, about post processing).To improve performance, you should combine the light intensity value with the light color (make sure that you convert the color to linear space first). If you are not using an HDR target, make sure that the combined intensity and color value is lower than one. This combined light intensity and color is stored in the second shader constant `DirLightColor`.

The material is defined by the following structure:

```
struct Material
{
    float3 normal;
    float4 diffuseColor;
    float specExp;
    float specIntensity;
};
```

The material values should be prepared in the pixel shader before calling the function that calculates the final lit color of the pixel. The normals should be in world space and normalized. The diffuse value can be a constant color or a sample from a texture. When a material doesn't support specular highlights, just set specExp to 1 and specIntensity to 0, otherwise use appropriate values based on the desired look (see explanation to specular light in the *How it works...* section of this recipe).

Here is the code for calculating the directional light value based on the input parameters:

```
float3 CalcDirectional(float3 position, Material material)
{
    // Phong diffuse
    float NDotL = dot(DirToLight, material.normal);
    float3 finalColor = DirLightColor.rgb * saturate(NDotL);

    // Blinn specular
    float3 ToEye = EyePosition.xyz - position;
    ToEye = normalize(ToEye);
    float3 HalfWay = normalize(ToEye + DirToLight);
    float NDotH = saturate(dot(HalfWay, material.normal));
    finalColor += DirLightColor.rgb * pow(NDotH, material.specExp) *
    material.specIntensity;

    return finalColor * material.diffuseColor.rgb;
}
```

This function takes the pixel's world position and material values, and it outputs the pixel's lit color value.

How it works...

The Blinn-Phong light equation used in the previous code is very popular, as it is easy to compute and provides pleasing visual results. The equation is split into two components: a diffuse and a specular component. The following figure shows the different vectors used in the directional light calculation:

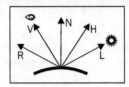

Diffuse light is defined as a light reflected by the mesh surface equally in all directions. As you can see from the calculation, the diffuse light value for a given pixel is only affected by the normal **N** and by the direction to light **L** using the dot product value. If you recall from linear algebra, the dot product equals to:

Dot(N, L) = |N||L|cos(α)

Where α is the angle between **N** and **L**. Since all vectors are normalized, the size of **N** and the size of **L** is one, so the dot product in this case is equal to the cosine of the angle between the vectors. This means that the diffuse light gets brighter, as the normal **N** and the direction to the light **L** get closer to being parallel and dimmer as they get closer to being perpendicular.

Specular light, as opposed to diffuse light, gets reflected by the mesh in a specific direction. Light coming from the light source gets reflected in the direction **R**. Calculating the reflection vector **R** is a bit expensive, so Blinn's equation provides a very good and fast approximation using the half-way vector **H** (the vector at half the angle between the direction to the viewer **V** and the direction to the light **L**). If you imagine how the **H** light is going to move when **V** and **L** move, you will see that the angle between **H** and **N** gets smaller when the angle between **R** and **V** gets smaller. Using the dot product of **N** and **H**, we get a good estimate to how close the view direction is to the reflected vector **R**.

The power function is then used to calculate the intensity of the reflected light for the given angle between **N** and **H**. The higher the material's specular exponent is, the smaller the light spread is.

There's more...

For performance reasons, it's very common to combine the ambient light calculation with the directional light in the same shader. In most scenes, there is only one directional light source, so by calculating both directional and ambient light in the same shader, you can save one draw call per mesh.

All you have to do is just add the value of the directional light to the ambient light value like this:

```
// Calculate the ambient color
float4 finalColor;
finalColor.rgb = CalcAmbient(Normal, material.diffuseColor.rgb);

// Calculate the directional light
finalColor.rgb += CalcDirectional(worldPosition, material);
```

Point light

Point light is a light source that emits light equally in all directions. A good example for cases where a point light should be used is for an exposed light bulb, lit torch, and any other light source that emits light evenly in all directions.

The following screenshot shows the bunny with a point light in front of its chest:

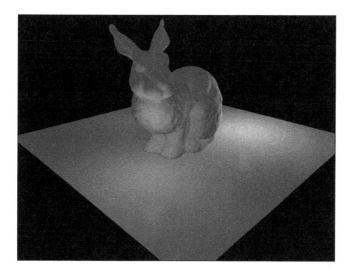

Looking at the previous screenshot, you may be wondering why you can't see the actual light source. With the exception of an ambient light, all the light sources featured in this chapter only calculate the first light bounce. Because we don't calculate the effect of rays hitting the camera directly from the light source, the light source is invisible. It is a common practice to render a mesh that represents the light source with a shader that outputs the light's color of multiplayer by its intensity. This type of shader is commonly known as an emissive shader.

Getting ready

Point lights extend the directional light calculation by making the direction between each pixel and the light source based on the pixel and light position (unlike the fixed direction used in directional light).

Instead of the direction value used by directional light, point lights use a position and the range values. The position should be the center of the light source. The range should be the edge of the point light's influence (the furthest distance light can travel from the source and affect the scene).

How to do it...

Similar to directional light, the point light is going to use the pixel position and the material structure. Remember that the normal has to be normalized and that the diffuse color has to be in linear space.

Instead of the direction vector used by directional light, point light requires a position in world space and a range in world space units. Inside the point light calculation, we need to divide the point lights range value. Since the GPU handles multiplication better than division, we store the Range value as *1/Range* (make sure that the range value is bigger than zero), so we can multiply instead of divide.

 1 / Range is called the reciprocal of Range.

We declare the position and reciprocal range inside the pixels header as follows:

```
cbuffer DirLightConstants : register( b0 )
{
    float3 PointLightPos   : packoffset( c0 );
    float PointLightRangeRcp  : packoffset( c0.w );
}
```

Here is the code for calculating the point light:

```
float3 CalcPoint(float3 position, Material material)
{
    float3 ToLight = PointLightPos.xyz - position;
    float3 ToEye = EyePosition.xyz - position;
    float DistToLight = length(ToLight);

    // Phong diffuse
```

```
    ToLight /= DistToLight; // Normalize
    float NDotL = saturate(dot(ToLight, material.normal));
    float3 finalColor = PointColor.rgb * NDotL;

    // Blinn specular
    ToEye = normalize(ToEye);
    float3 HalfWay = normalize(ToEye + ToLight);
    float NDotH = saturate(dot(HalfWay, material.normal));
    finalColor += PointColor.rgb * pow(NDotH, material.specExp) *
material.specIntensity;

    // Attenuation
    float DistToLightNorm = 1.0 - saturate(DistToLight *
PointLightRangeRcp);
    float Attn = DistToLightNorm * DistToLightNorm;
    finalColor *= material.diffuseColor * Attn;

    return finalColor;
}
```

This function takes the pixel's world position and material values, and outputs the pixel's lit color value.

How it works...

As with the directional light, the Blinn-Phong model is used for point light calculation. The main difference is that the light direction is no longer constant for all the pixels. Since the point light emits light in a sphere pattern, the light direction is calculated per pixel as the normalized vector from the pixel position to the light source position.

The attenuation calculation fades the light based on distance from the source. In the featured code, a squared attenuation is used. Depending on the desired look, you may find a different function more suitable.

 You can get a different attenuation value for each light source by using the HLSL pow function with a per-light source term.

Spot light

Spot light is a light source that emits light from a given position in a cone shape that is rounded at its base. The following screenshot shows a spot light pointed at the bunny's head:

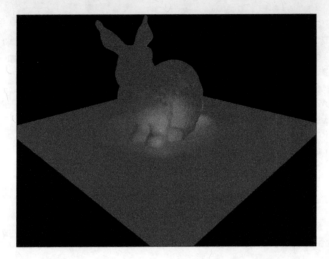

The cone shape of the spot light is perfect for representing flash lights, vehicle's front lights, and other lights that are focused in a specific direction.

Getting ready

In addition to all the values needed for point light sources, a spot light has a direction and two angles to represent its cone. The two cone angles split the cone into an inner cone, where light intensity is even, and an outer cone, where light attenuates if it gets closer to the cone's border. The following screenshot shows the spot light direction as **D**, the inner to outer cone angle as **α**, and the outer cone angle as **θ**:

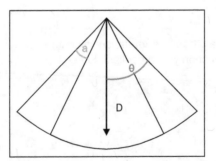

Unlike the point light, where light intensity attenuates only over distance, spot lights intensity also attenuates across the angle **α**. When a light ray angle from the center is inside the range of **α**, the light gets dimmer; the dimmer the light, the closer the angle is to **θ**.

How to do it...

For the spot light calculation, we will need all the values used for point light sources plus the additional three values mentioned in the previous section. The following deceleration contains the previous and new values:

```
cbuffer SpotLightConstants : register( b0 )
{
    float3 SpotLightPos        : packoffset( c0 );
    float SpotLightRangeRcp    : packoffset( c0.w );
    float3 SpotLightDir        : packoffset( c1 );
    float SpotCosOuterCone     : packoffset( c1.w );
    float SpotInnerConeRcp     : packoffset( c2 );
}
```

Like the directional light's direction, the spot light's direction has to be normalized and inverted, so it would point to the light (just pass it to the shader, minus the light direction). The inverted direction is stored in the shader constant `SpotLightDir`.

Reciprocal range is stored in the shader constant `SpotLightRangeRcp`.

When picking the inner and outer cone angles, always make sure that the outer angle is bigger than the outer to inner angle. During the spot light calculation, we will be using the cosine of the inner and outer angles. Calculating the cosine values over and over for every lit pixel in the pixel shader is bad for performance. We avoid this overhead by calculating the cosine values on the CPU and passing them to the GPU. The two angle cosine values are stored in the shader constants `SpotCosOuterCone` and `SpotCosInnerCone`.

The code to calculate the spot light is very similar to the point light code:

```
float3 CalcSpot(float3 position, Material material)
{
    float3 ToLight = SpotLightPos - position;
    float3 ToEye = EyePosition.xyz - position;
    float DistToLight = length(ToLight);

    // Phong diffuse
    ToLight /= DistToLight; // Normalize
    float NDotL = saturate(dot(ToLight, material.normal));
    float3 finalColor = SpotColor.rgb * NDotL;

    // Blinn specular
```

```
ToEye = normalize(ToEye);
float3 HalfWay = normalize(ToEye + ToLight);
float NDotH = saturate(dot(HalfWay, material.normal));
finalColor += SpotColor.rgb * pow(NDotH, material.specExp) *
material.specIntensity;

// Cone attenuation
float conAtt = saturate((cosAng - SpotCosOuterCone) *
SpotCosInnerConeRcp);
conAtt *= conAtt;

// Attenuation
float DistToLightNorm = 1.0 - saturate(DistToLight *
SpotLightRangeRcp);
float Attn = DistToLightNorm * DistToLightNorm;
finalColor *= material.diffuseColor * Attn * conAtt;

return finalColor;
}
```

As with the previous two light functions, this function takes the pixel's world position and material values, and outputs the pixel's lit color value.

How it works...

As with the previous light sources, the spot light is using the Blinn-Phong model. The only difference in the code is the cone attenuation, which gets combined with the distance attenuation. To account for the cone shape, we first have to find the angle between the pixel to light vector and the light vector. For that calculation we use the dot product and get the cosine of that angle. We then subtract the cosine of the outer cone angle from that value and end up with one of the three optional results:

- ▶ If the result is higher than the cosine of the inner cone, we will get a value of 1 and the light affect will be fully on

- ▶ If the result is lower than the cosine of the inner cone but higher than zero, the pixel is inside the attenuation range and the light will get dimmer based on the size of the angle

- ▶ If the result is lower than zero, the pixel was outside the range of the outer cone and the light will not affect the pixel

Capsule light

Capsule light, as the name implies, is a light source shaped as a capsule. Unlike spot and point light sources that have a point origin, the capsule light source has a line at its origin and it is emitting light evenly in all directions. The following screenshot shows a red capsule light source:

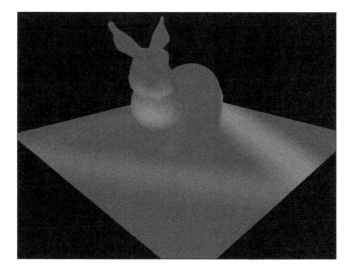

Capsule lights can be used to represent fluorescent tubes or a lightsaber.

Getting ready

Capsules can be thought of as a sphere split into two halves, which are then extruded by the length of the capsule light's line length. The following figure shows the line start point **A** and end points **B** and **R** are the light's range:

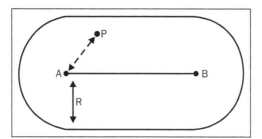

How to do it...

Capsule lights use the following constant buffer in their pixel shader:

```
cbuffer CapsuleLightConstants : register( b0 )
{
  float3 CapsuleLightPos      : packoffset( c0 );
  float CapsuleLightRangeRcp  : packoffset( c0.w );
  float3 CapsuleLightDir      : packoffset( c1 );
  float CapsuleLightLen       : packoffset( c1.w );
  float3 CapsuleLightColor    : packoffset( c2 );
}
```

Point **A**, referred to as the starting point is stored in the shader constant `CapsuleLightPos`.

In order to keep the math simple, instead of using the end point directly, we are going to use the normalized direction from **A** to **B** and the line's length (distance from point **A** to point **B**). We store the capsule's direction in the constant `CapsuleLightDir` and the length in `CapsuleLightLen`.

Similar to the point and spot lights, we store the range.

The code for calculating the capsule light looks like this:

```
float3 CalcCapsule(float3 position, Material material)
{
    float3 ToEye = EyePosition.xyz - position;

    // Find the shortest distance between the pixel and capsules segment
    float3 ToCapsuleStart = position - CapsuleLightPos;
    float DistOnLine = dot(ToCapsuleStart, CapsuleDirLen.xyz) /
    CapsuleLightRange;
    DistOnLine = saturate(DistOnLine) * CapsuleLightRange;
    float3 PointOnLine = CapsuleLightPos + CapsuleLightDir * DistOnLine;
    float3 ToLight = PointOnLine - position;
    float DistToLight = length(ToLight);

    // Phong diffuse
    ToLight /= DistToLight; // Normalize
    float NDotL = saturate(dot(ToLight, material.normal));
    float3 finalColor = material.diffuseColor * NDotL;

    // Blinn specular
    ToEye = normalize(ToEye);
    float3 HalfWay = normalize(ToEye + ToLight);
```

```
    float NDotH = saturate(dot(HalfWay, material.normal));
    finalColor += pow(NDotH, material.specExp) * material.specIntensity;

    // Attenuation
    float DistToLightNorm = 1.0 - saturate(DistToLight *
    CapsuleLightRangeRcp);
    float Attn = DistToLightNorm * DistToLightNorm;
    finalColor *= CapsuleLightColor.rgb * CapsuleIntensity * Attn;

    return finalColor;
}
```

This function takes the pixel's world position and material values, and it outputs the pixel's lit color value.

How it works...

Look closely at the code and you will notice that it's basically the point light code except for the pixel to light position vector calculation. The idea is to find the closest point on the line to the pixel position. Once found, the vector to light is calculated by subtracting the closest position from the pixel position.

Finding the closest point on the line is done using the dot product. If you recall, the dot product result is the projected length of one vector over the other. By calculating the dot product of the vector AP with the capsule direction, we find the distance on the line from **A** to the closest point. We then have three possible results:

▸ The value is negative (outside the line from A's side); in this case the closest point is **A**

▸ The value is positive, but it's bigger than the line's length (outside the line from **B**'s side); in this case the closest point is **B**

▸ The value is within the line's length and it doesn't need any modifications

HLSL is not very good with code branches, so instead of using `if` statements, the value found is normalized by dividing with the line's length and using the saturate instruction (clamp the value to zero and one). This affectively takes care of situations one and two. By multiplying the normalized value with the line's length, we end up with the correct distance from **A**. Now we can find the closest point by adding **A** and the distance of the capsule direction.

From that point on, the calculations are exactly the same as the point lights.

Projected texture – point light

All light sources covered up to this point spread light in an even intensity distribution. However, sometimes a light source has a more sophisticated intensity distribution pattern. For example, a lamp shade can change the light intensity distribution and make it uneven. A different situation is when the intensity is even, but the color isn't due to something covering the light source. Using math to represent those and other situations may be too complex or have a negative effect on rendering performance. The most common solution in these cases is to use a texture that represents the intensity/color pattern emitted by these light sources.

The following screenshot shows a point light source projecting a texture with stars on the bunny:

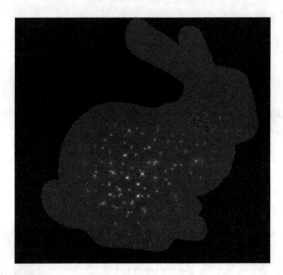

Getting ready

To project a texture with a point light, you will need a texture that wraps around the point light's center. The best option is to use a cube map texture. The cube map texture is a group of six 2D textures that cover the faces of an imaginary box. Microsoft's Direct X SDK comes with a texture tool called DirectX Texture Tool, which helps you group six 2D textures and store them in an DXT format.

In order to sample the cube map texture, you will need a direction vector that points from the light source in the pixel directions. When the light is stationary, the texture can be prepared so the vector is the world-space direction from the light center to the pixel. If the light can rotate, the sampling direction has to take the rotation into an account. In those cases, you will need a matrix that transforms a direction in world space into the light's space.

Sampling the cube map texture will require a linear sampler state. Fill a D3D11_SAMPLER_DESC object with the following values:

Sampler State Descriptor Parameter	Value
Fliter	D3D11_FILTER_MIN_MAG_MIP_LINEAR
AddressU	D3D11_TEXTURE_ADDRESS_WRAP
AddressV	D3D11_TEXTURE_ADDRESS_WRAP
AddressW	D3D11_TEXTURE_ADDRESS_WRAP
MaxAnisotropy	1
ComparisonFunc	D3D11_COMPARISON_ALWAYS
MaxLOD	D3D11_FLOAT32_MAX

The reset of the descriptor fields should be set to 0.

Create the actual sampler state from the descriptor using the device function CreateSamplerState.

How to do it...

To keep the code generic, we are going to support light source rotation. For light sources that don't rotate, just use an identity matrix. For performance reasons, it is preferred to calculate the sampling direction in the vertex shader using the vertex world position.

In order to transform a position into light space, we are going to use the following shader constant matrix:

```
float4x4 LightTransform;
```

Usually when a light rotates, it is attached to some scene model that represents the light source. The model has a transformation from the model space to world space. All you need to do is inverse that transformation and use it as LightTransform.

Computing the sampling direction can be done in either the vertex or pixel shader using the following code (again, it is recommended to do this in the vertex shader):

```
float3 GetDirToLight(float3 WorldPosition)
{
    float3 ToLgiht = LightTransform[3].xyz + WorldPosition;
    return mul(ToLgiht.xyz, (float3x3)LightTransform);
}
```

This function takes the vertex/pixel position as argument and returns the sampling direction. If you choose to calculate the sampling direction in the vertex shader, make sure that you pass the result to the pixel shader.

In addition to the cube map texture, we will need a single intensity value. If you recall from the basic point light implementation, the intensity used to be combined with the color value. Now that the color value is sampled from a texture, it can no longer be combined with the intensity on the CPU. The intensity is stored in the following global variable:

```
float PointIntensity;
```

We will be accessing the cube map texture in the pixel shader using the following shader resource view deceleration:

```
TextureCube ProjLightTex : register( t0 );
```

As mentioned in the *Getting ready* section of this recipe, sampling the cube map will also require a linear sampler. Add the following sampler state deceleration in your pixel shader:

```
SamplerState LinearSampler : register( s0 );
```

Using the sampling direction, we can now find the per-pixel color value of the light using the following code:

```
float3 GetLightColor(float3 SampleDirection)
{
    return PointIntensity * ProjLightTex.Sample(LinearSampler,
    SampleDirection);
}
```

The returned color intensity should now be used to calculate color affect using the following code:

```
float3 CalcPoint(float3 LightColor, float3 position, Material
material)
{
    float3 ToLight = PointLightPos - position;
    float3 ToEye = EyePosition.xyz - position;
    float DistToLight = length(ToLight);

    // Phong diffuse
    ToLight /= DistToLight; // Normalize
    float NDotL = saturate(dot(ToLight, material.normal));
    float3 finalColor = LightColor * NDotL;

    // Blinn specular
    ToEye = normalize(ToEye);
    float3 HalfWay = normalize(ToEye + ToLight);
    float NDotH = saturate(dot(HalfWay, material.normal));
    finalColor += LightColor * pow(NDotH, material.specExp) *
    material.specIntensity;
```

```
    // Attenuation
    float DistToLightNorm = 1.0 - saturate(DistToLight *
    PointLightRangeRcp);
    float Attn = DistToLightNorm * DistToLightNorm;
    finalColor *= material.diffuseColor * Attn;

    return finalColor;
}
```

If you compare this code with the code used for spot lights that don't use projected textures, you will notice that the only difference is the light color getting passed as an argument instead of the constant color used in the basic implementation.

How it works...

We consider the cube map texture to be in light space. This means that the values in the texture stay constant regardless of the light rotation and movement. By transforming the direction to the light into light space, we can sample the texture with it and get the color value. Aside from the sampled color value replacing the constant color from the basic point light implementation, the code stays exactly the same.

Projected texture – spot light

Similar to point lights with projected textures, spot lights can use a 2D texture instead of a constant color value. The following screenshot shows a spot light projecting a rainbow pattern on the bunny:

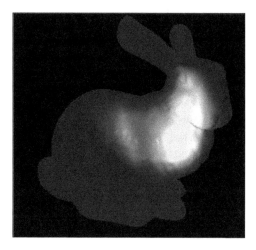

Getting ready

Due to the cone shape of the spot light, there is no point in using a cube map texture. Most spot light sources use a cone opening angle of 90 degrees or less, which is equivalent to a single cube map face. This makes using the cube map a waste of memory in this case. Instead, we will be using a 2D texture.

Projecting a 2D texture is a little more complicated compared to the point light. In addition to the transformation from world space to light space, we will need a projection matrix. For performance reasons, those two matrices should be combined to a single matrix by multiplying them in the following order:

*FinalMatrix = ToLightSpaceMatrix * LightProjectionMatrix*

Generating this final matrix is similar to how the matrices used for rendering the scene get generated. If you have a system that handles the conversion of camera information into matrices, you may benefit from defining a camera for the spot light, so you can easily get the appropriate matrices.

How to do it...

Spot light projection matrix can be calculated in the same way the projection matrix is calculated for the scene's camera. If you are unfamiliar with how this matrix is calculated, just use the following formulas:

$$
\begin{bmatrix} w & 0 & 0 & 0 \\ 0 & h & 0 & 0 \\ 0 & 0 & Q & 0 \\ 0 & 0 & -QZ_n & 0 \end{bmatrix}
\qquad
\begin{aligned}
w &= \cot\left(\frac{fov_w}{2}\right) \\[2mm]
h &= \cot\left(\frac{fov_h}{2}\right) \\[2mm]
Q &= \frac{Z_f}{Z_f - Z_n}
\end{aligned}
$$

In our case, both *w* and *h* are equal to the cotangent of the outer cone angle. Zfar is just the range of the light source. Znear was not used in the previous implementation and it should be set to a relatively small value (when we go over shadow mapping, this value's meaning will become clear). For now just use the lights range times 0.0001 as Znear's value.

The combined matrix should be stored to the vertex shader constant:

```
float4x4 LightViewProjection;
```

Getting the texture coordinate from the combined matrix is handled by the following code:

```
float2 GetProjPos(float4 WorldPosition)
{
    float3 ProjTexXYW = mul(WorldPosition, LightViewProjection).xyw;
    ProjTexXYW.xy /= ProjTexXYW.z; // Perspective correction
    Float2 UV = (ProjTexXYW.xy + 1.0.xx) * float2(0.5, -0.5); //
    Convert to normalized space
    return UV;
}
```

This function takes the world position as four components (w should be set to 1) as parameter and returns the projected texture UV sampling coordinates. This function should be called in the vertex shader.

The texture coordinates should be than passed to the pixel shader, so they can be used for sampling the texture. In order to sample the texture, the following shader resource view should be defined in the pixel shader:

```
Texture2D ProjLightTex : register( t0 );
```

Sampling the texture is done in the pixel shader with the following code:

```
float3 GetLightColor(float2 UV)
{
    return SpotIntensity * ProjLightTex.Sample(LinearSampler, UV);
}
```

This function takes the UV coordinates as parameter and returns the light's color intensity for the pixel. Similar to point lights with projected textures, the color sampled from the texture is then multiplied by the intensity to get the color intensity value used in the lighting code.

The only thing left to do is to use the light color intensity and light the mesh. This is handled by the following code:

```
float3 CalcSpot(float3 LightColor, float3 position, Material material)
{
    float3 ToLight = SpotLightPos - position;
    float3 ToEye = EyePosition.xyz - position;
    float DistToLight = length(ToLight);

    // Phong diffuse
    ToLight /= DistToLight; // Normalize
    float NDotL = saturate(dot(ToLight, material.normal));
    float3 finalColor = LightColor * NDotL;

    // Blinn specular
```

```
    ToEye = normalize(ToEye);
    float3 HalfWay = normalize(ToEye + ToLight);
    float NDotH = saturate(dot(HalfWay, material.normal));
    finalColor += LightColor * pow(NDotH, material.specExp) *
    material.specIntensity;

    // Cone attenuation
    float cosAng = dot(SpotLightDir, ToLight);
    float conAtt = saturate((cosAng - SpotCosOuterCone) /
    SpotCosInnerCone);
    conAtt *= conAtt;

    // Attenuation
    float DistToLightNorm = 1.0 - saturate(DistToLight /
    SpotLightRange);
    float Attn = DistToLightNorm * DistToLightNorm;
    finalColor *= material.diffuseColor * Attn * conAtt;

    return finalColor;
}
```

Similar to the point light implementation with projected texture support, you will notice that the only difference compared to the basic spot light code is the light color getting passed as an argument.

How it works...

Converting the world space position to texture coordinates is very similar to the way world positions get converted to the screen's clip space by the GPU. After multiplying the world position with the combined matrix, the position gets converted to projected space that can be then converted to clip space (X and Y values that are inside the lights influence will have the value -1 to 1). We then normalize the clip space values (X and Y range becomes 0 to 1), which are the UV range we need texture sampling for.

All values passed to the vertex shader are linearly interpolated, so the values passed from the vertex shader to the pixel shader will be interpolated correctly for each pixel based on the UV values calculated for the three vertices, which make the triangle the pixel was rasterized from.

Multiple light types in a single pass

In this chapter's introduction, we saw that the performance can be optimized by combining multiple lights into a single draw call. The GPU registers are big enough to contain four float values. We can take advantage of that size and calculate four lights at a time.

Unfortunately, one drawback of this approach is the lack of support for projected textures. One way around this issue is to render light sources that use projected textures separately from the rest of the light sources. This limit may not be all that bad depending on the rendered scene light setup.

Getting ready

Unlike the previous examples, this time you will have to send light data to the GPU in groups of four. Since it's not likely that all four light sources are going to be of the same type, the code handles all three light types in a generic way. In cases where the drawn mesh is affected by less than four lights, you can always disable lights by turning their color to fully black.

How to do it...

In order to take full advantage of the GPU's vectorized math operations, all the light source values are going to be packed in groups of four. Here is a simple illustration that explains how four-three component variables can be packed into three-four component variables:

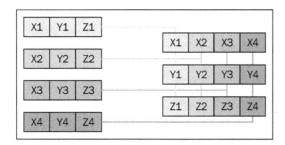

All variable packing should be done on the CPU. Keeping in mind that the constant registers of the GPU are the size of four floats, this packing is more efficient compared to the single light version, where most of the values use only three floats and waste the forth one.

Light positions X, Y, and Z components of each of the four light sources are packed into the following shader constants:

```
float4 LightPosX;
float4 LightPosY;
float4 LightPosZ;
```

Light directions are separated to X, Y, and Z components as well. This group of constants is used for both, spot and capsule light source directions. For point lights make sure to set the respected value in each constant to 0:

```
float4 LightDirX;
float4 LightDirY;
float4 LightDirZ;
```

Light color is separated to R, G, and B components. For disabled light sources just set the respected values to 0:

```
float4 LightColorR;
float4 LightColorG;
float4 LightColorB;
```

As before, you should combine the color and intensity of each light before passing the values to the GPU.

All four light ranges are stored in a single four-component constant:

```
float4 LightRange;
```

All four lights' capsule lengths are stored in a single four-component constant. For noncapsule lights just store the respected value to 0:

```
float4 CapsuleLen;
```

Spot light's cosine outer cone angle is again stored in a four-component constant. For nonspot light sources set the respected value to -2:

```
float4 SpotCosOuterCone;
```

Unlike the single spot light, for the inner cone angle we are going to store one over the spot light's cosine inner cone angle. For nonspot light sources set the respected value to 1:

```
float4 SpotCosInnerConeRcp;
```

We are going to use two new helper functions that will help us calculate the dot product of four component vectors. The first one calculates the dot product between two groups of three-four component variable. The return value is a four-component variable with the four-dot product values. The code is as follows:

```
float4 dot4x4(float4 aX, float4 aY, float4 aZ, float4 bX, float4 bY,
float4 bZ)
{
    return aX * bX + aY * bY + aZ * bZ;
}
```

The second helper function calculates the dot product of three-four component variables with a single three-component variable:

```
float4 dot4x1(float4 aX, float4 aY, float4 aZ, float3 b)
{
    return aX * b.xxxx + aY * b.yyyy + aZ * b.zzzz;
}
```

Finally, the code to calculate the lighting for the four light sources is as follows:

```
float3 ToEye = EyePosition.xyz - position;

// Find the shortest distance between the pixel and capsules segment
float4 ToCapsuleStartX = position.xxxx - LightPosX;
float4 ToCapsuleStartY = position.yyyy - LightPosY;
float4 ToCapsuleStartZ = position.zzzz - LightPosZ;
float4 DistOnLine = dot4x4(ToCapsuleStartX, ToCapsuleStartY,
ToCapsuleStartZ, LightDirX, LightDirY, LightDirZ);
float4 CapsuleLenSafe = max(CapsuleLen, 1.e-6);
DistOnLine = CapsuleLen * saturate(DistOnLine / CapsuleLenSafe);
float4 PointOnLineX = LightPosX + LightDirX * DistOnLine;
float4 PointOnLineY = LightPosY + LightDirY * DistOnLine;
float4 PointOnLineZ = LightPosZ + LightDirZ * DistOnLine;
float4 ToLightX = PointOnLineX - position.xxxx;
float4 ToLightY = PointOnLineY - position.yyyy;
float4 ToLightZ = PointOnLineZ - position.zzzz;
float4 DistToLightSqr = dot4x4(ToLightX, ToLightY, ToLightZ,
ToLightX, ToLightY, ToLightZ);
float4 DistToLight = sqrt(DistToLightSqr);

// Phong diffuse
ToLightX /= DistToLight; // Normalize
ToLightY /= DistToLight; // Normalize
ToLightZ /= DistToLight; // Normalize
float4 NDotL = saturate(dot4x1(ToLightX, ToLightY, ToLightZ,
material.normal));
//float3 finalColor = float3(dot(LightColorR, NDotL),
dot(LightColorG, NDotL), dot(LightColorB, NDotL));

// Blinn specular
ToEye = normalize(ToEye);
float4 HalfWayX = ToEye.xxxx + ToLightX;
float4 HalfWayY = ToEye.yyyy + ToLightY;
float4 HalfWayZ = ToEye.zzzz + ToLightZ;
float4 HalfWaySize = sqrt(dot4x4(HalfWayX, HalfWayY, HalfWayZ,
HalfWayX, HalfWayY, HalfWayZ));
```

```
float4 NDotH = saturate(dot4x1(HalfWayX / HalfWaySize, HalfWayY /
HalfWaySize, HalfWayZ / HalfWaySize, material.normal));
float4 SpecValue = pow(NDotH, material.specExp.xxxx) *
material.specIntensity;
//finalColor += float3(dot(LightColorR, SpecValue),
dot(LightColorG, SpecValue), dot(LightColorB, SpecValue));

// Cone attenuation
float4 cosAng = dot4x4(LightDirX, LightDirY, LightDirZ, ToLightX,
ToLightY, ToLightZ);
float4 conAtt = saturate((cosAng - SpotCosOuterCone) *
SpotCosInnerConeRcp);
conAtt *= conAtt;

// Attenuation
float4 DistToLightNorm = 1.0 - saturate(DistToLight *
LightRangeRcp);
float4 Attn = DistToLightNorm * DistToLightNorm;
Attn *= conAtt; // Include the cone attenuation

// Calculate the final color value
float4 pixelIntensity = (NDotL + SpecValue) * Attn;
float3 finalColor = float3(dot(LightColorR, pixelIntensity),
dot(LightColorG, pixelIntensity), dot(LightColorB, pixelIntensity));
finalColor *= material.diffuseColor;

return finalColor;
```

How it works...

Though this code is longer than the one used in the previous recipes, it basically works in the exact same way as in the single light source case. In order to support the three different light sources in a single code path, both the capsule light's closest point to line and the spot lights cone attenuation are used together.

If you compare the single light version of the code with the multiple lights version, you will notice that all the operations are done in the exact same order. The only change is that each operation that uses the packed constants has to be executed three times and the result has to be combined into a single four-component vector.

There's more...

Don't think that you are limited to four lights at a time just because of the GPU's constant size. You can rewrite `CalcFourLights` to take the light constant parameters as inputs, so you could call this function more than once in a shader.

Some scenes don't use all three light types. You can remove either the spot or capsule light support if those are not needed (point lights are at the base of the code, so those have to be supported). This will reduce the shader size and improve performance.

Another possible optimization is to combine the ambient, directional, and multiple lights code into a single shader. This will reduce the total amount of draw calls needed and will improve performance.

2

Deferred Shading

In this chapter we will cover:

- ▶ GBuffer generation
- ▶ GBuffer unpacking
- ▶ Directional light
- ▶ Point light
- ▶ Capsule light
- ▶ Spot light

Introduction

In the previous chapter, we mentioned that forward lighting has performance limitations. If you recall the basic algorithm for forward lighting, we potentially need to draw every mesh once for each light source. For a scene with N light sources and M meshes, we may need N times M draw calls. This means that every additional mesh is going to add up to N new draw calls and each new light source will add up to M new draw calls. Those numbers add up very quickly and pose a major limitation on the amount of elements you can render in an interactive frame rate.

Over the years, as graphic cards' memory size increased and GPUs commonly supported rendering to multiple render targets (MRT), a couple of solutions to the performance problems were developed. At their core, all the solutions work on the same principle – separate the scene rendering from the lighting. This means that when using those new methods, a scene with M meshes and N light sources is going to only take M + N draw calls. That is obviously much better than the performance offered by forward lighting.

One thing that is common to all these methods is the **Geometry Buffer**, or **GBuffer** for short. A GBuffer is a collection of render targets with the same pixel count as the final image that get filled with the scene per-pixel information. All the render targets in the GBuffer have the same size due to hardware limitations. The bare minimum for a GBuffer is to contain the scenes' per-pixel depth and normal.

As an example, let's look at the two of the most popular variations of those methods. The first one is called deferred lighting. Deferred lighting works in three stages:

  ▸ Render the scene into a GBuffer that contains depth and normal

  ▸ For each light source, find the affected pixels, read the corresponding GBuffer data, calculate the light value and store that value into a light accumulation render target

  ▸ Render the scene a second time and for each pixel combine the accumulated light with the mesh color to get the final pixel color

Although we render the scene twice when using deferred lighting, the scene from light separation is much better compared to forward. These two qualities made deferred lighting a very attractive option on the Xbox 360 (small GBuffer means no need to tile in EDRAM). However, if you are developing on a modern PC GPU (or a next gen consoles), you probably won't have to worry about available GPU memory as much. This brings us to the second method – deferred shading.

Deferred shading is not much different compared to deferred lighting. It works in two stages:

  ▸ Render the scene into a GBuffer that contains everything needed for full shading and post processing. This includes a minimum of depth, normal, per-pixel mesh color, specular power, and specular intensity

  ▸ For each light source, find the affected pixels, read the corresponding GBuffer data, calculate the pixels lit color, and store it to an accumulation buffer

Two things changed compared to deferred lighting. First thing, the GBuffer has to contain more data so we can calculate each pixels lit and shaded color. Second change is that the scene only has to be rendered once.

Taking into account the large amount of memory available on new graphic cards, the only argument in favor of deferred lighting is the memory bandwidth performance. When rendering, the memory bandwidth relates to the amount of memory needed per pixel. The more memory each pixel requires, the slower it becomes to read/write the pixel information from memory. In my opinion, this advantage is neglectable on modern GPUs.

On the other hand, deferred shading is simpler compared to deferred lighting and requires less shaders. In addition, the extra data stored in the GBuffer is very useful for post processing. Because of these considerations, this chapter is going to focus on deferred shading.

If you do choose to go with deferred lighting, converting the HLSL code presented in this chapter's recipes should be relatively easy and can be a good exercise to help you understand the difference between the two techniques.

Before we continue it's important to explain the biggest limitation of these deferred methods. As the GBuffer can only contain a single set of data for each pixel in the scene, those deferred methods can only be used for fully opaque scene elements. Although this sounds like a big restriction, it doesn't mean you can't use these techniques that contain translucent elements. A common solution to this restriction is to light the opaque elements in the scene with a deferred method, than use forward rendering to add the semi-transparent elements on top. If you recall how the full forward rendering solution works this "opaque first, semi-transparent later" order was used there as well.

GBuffer generation

Unlike forward lighting, where we just started rendering the scene with the lights, deferred shading requires the GBuffer to be generated before any light related calculation can take place.

Choosing the structure of the GBuffer is a very important decision. The bare minimum you can get away with is storing depth, base color, normal, specular power, and specular intensity. For this configuration, you will need between three to four render targets.

Storing depth and base color is straight forward. Use a D24S8 render target for the depth and stencil. For base color we will be using an A8R8G8B8 render target as the base color values are usually sampled from 8 bit per-pixel texture. The alpha channel of the color target can be used to store the specular intensity. Storing normals on the other hand is not that simple and requires some planning. To keep this section relatively short, we are only going to cover three storage methods that have been used in popular AAA games.

 Note that normalized normals don't use the full range of the XYZ components (for example, the combination <1, 1, 0> is never going to be used as it will be normalized to <0.7071, 0.7071, 0>). This makes the normal precision loss worse, compared to values that use the full precision range, when we write the normal into a GBuffer target that uses 8 bit or 16 bit per channel.

The most straight forward storage method is to encode the normalized normal range [-1, 1] into the range [0, 1] than just write it to three channels of a render target. During the days of DirectX9, the only available render target formats were 32 bit per pixel (remember the limitation that all render targets have to be the same size) that had 8 bits per channel. Each normal channel coming from the Vertex shader can contain between 2 to 4 bytes of data (depending on use of half or float precision). Using this method we will store those normals into 1 byte per channel, so a lot of precision is going to get lost in the process.

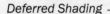

Pros: This method is the easiest to implement and requires the least math to encode and decode. It supports alpha blending. The unused alpha channel can be used for storing the encoded specular power.

Cons: Low quality normals after decoding and a poor use of the 24 bit of memory used by this method.

Reducing the normals precision due to storage limitations can lead to what is known as bending, especially when calculating specular lighting. Bending usually shows up as a non-smooth color. This brings us to the two other methods that use two 16 bit per channel format to store the normals.

The AAA title Killzone 2 stored the XY component of each normal in view space. When the normal is needed, the Z component is reconstructed from the other two components. While encoding the XY components, the sign of the Z component has to be encoded as well (normals can point away from the camera when perspective is used).

Pros: Better quality compared to the previous method due to the increased per-channel memory size.

Cons: If you choose to calculate lighting in world space, this method can become heavy on the decoding math. Also, normals have to be decoded before performing any math operations on them, so alpha blending of the encoded normals is out of the question.

The final method was used in the game Crysis 2 by Crytek. This method uses the same math normally used when a 3D vector is converted to a 2D UV that is used for sampling paraboloid textures. Packing into the GBuffer works by normalizing the XY components of the normal (causing their size to equal one) and multiplied by normalized Z component (Z component range is transformed from [-1, 1] to [0, 1]). Unpacking is done by reversing the math.

Pros: Good quality and can store normals in world space. Supports alpha blending.

Cons: Math heavy.

These days, DirectX11 offers a new render target format which stores three channels using the full range of a 32 bit render target. This new format separates those 32 bits by using 11 bits for the first two channels and 10 bit for the last channel. By using this storage format we can use the first method offered, only this time the precision loss is not going to be as bad. My experience with this format was very good, so I am going to use this method for the remainder of this chapter.

Last thing to consider is the specular power storage. There is a good chance you will be using only a very limited range of values for specular power. In most game related cases you can get away by using a single 8 bit channel of a forth render target to store the normalized specular power value.

Getting ready

First thing you will need is to allocate the GBuffers render targets. As described in the preceding section, the layout we will use for this recipe is shown in the following illustration:

24-Bit Depth	8-Bit Stencil
24-Bit Base Color RGB	8-Bit Specular Intensity
32-Bit Normals	
8-Bit Specular Power	? ? ?

The format types we are going to use for each one of these render targets is as follows:

Render target number	Use	Format
0	Depth and stencil	Texture: `R24G8_TYPELESS` Depth stencil view: `D24_UNORM_S8_UINT` Resource view: `R24_UNORM_X8_TYPELESS`
1	Base color, specular intensity	Texture: `R8G8B8A8_UNORM` Resource view: Same
2	Normal	Texture: `R11G11B10_FLOAT` Resource view: Same
3	Specular power	Texture: `R8G8B8A8_UNORM` Resource view: Same

The following image visualizes this setup with the GBuffer render targets rendered at the bottom from the depth buffer on the left to the specular value on the right:

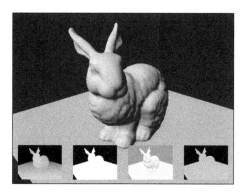

It is recommended by the GPU vendors to always allocate the GBuffer render targets first and in sequence. That way the graphics cards driver assumes those targets are important and will be used together.

 Unless you are using Multisample Anti-Aliasing MSAA, it's very likely that you would want your GBuffer textures to have the exact same size as your application / games back buffer. If that's the case, it is recommended that you save memory by using the back-buffer depth render target in your GBuffer instead of allocating a new one.

How to do it...

Filling the GBuffer with data is pretty straightforward. Each frame you start by clearing the depth view with depth equal to 1 and the other targets to black as you would normally do. Then you need to set all four targets in the order presented by the diagram (see the *Getting started...* section). Once the targets are ready, it's time to render the opaque elements of the scene.

Unlike the shaders used for forward rendering, this time we are going to output three color values instead of one. This is done using the following shader output structure:

```
struct PS_GBUFFER_OUT
{
    float4ColorSpecInt   : SV_TARGET0;
    float4 Normal    : SV_TARGET1;
    float4SpecPow    : SV_TARGET2;
};
```

As you can see, this output structure outputs all the values we want to write into the GBuffer in the same order the GBuffer views are set for rendering. Note that the normal will be stored as a four component value instead of three. This is due to a compiler restriction that won't allow a pixel shader to output any other type of floating point value. Just fill the W component with something (it will be ignored when written into the target).

Normalizing the specular power is straightforward. Assuming the range is constant you can write the minimum value and the affective range (maximum - minimum) value into shader constants like this:

```
staticconst float2 g_SpecPowerRange = { 0.1, 250.0 };
```

Finally, we need to pack all the different values into the GBuffer format. This is handled by the following function:

```
PS_GBUFFER_OUT PackGBuffer(float3 BaseColor, float3 Normal, float
SpecIntensity, float SpecPower)
{
    PS_GBUFFER_OUT Out;

    // Normalize the specular power
    floatSpecPowerNorm = (SpecPower - g_SpecPowerRange.x) /
    g_SpecPowerRange.y;

    // Pack all the data into the GBuffer structure
    Out.ColorSpecInt = float4(BaseColor.rgb, SpecIntensity);
    Out.Normal = float4(Normal.xyz * 0.5 + 0.5, 0.0);
    Out.SpecPow = float4(SpecPowerNorm, 0.0, 0.0, 0.0);

    return Out;
}
```

This function takes the base color, normalized normal, specular intensity, and specular power as input and outputs those values packed in the GBuffer format.

A typical pixel shader will look similar to this:

```
PS_GBUFFER_OUT DrawPS( VS_OUTPUT In )
{
    float3DiffuseColor;
    float3 Normal;
    floatSpecIntensity;
    floatSpecPower;

    ...

    // Fill all the values from vertex shader input or from texture
    samples

    ...

    returnPackGBuffer(DiffuseColor, normalize(Normal), SpecIntensity,
    SpecPower);
}
```

Once you render all the opaque scene elements into the GBuffer you should unset all the views in preparation to the lighting part.

How it works...

Most of the packing process should be self-explanatory. It is important to normalize the normal before passing it to the `PackGBuffer` function. Failing to do so can result in unneeded loss of precision.

Note that due to the specular power encoding, we no longer have to worry about zero or negative specular power. Once we unpack the GBuffer, a value of zero will be mapped to the minimum value we encoded with.

Depending on the specular power values and the range you plan on using, you may want to consider using a non-linear encoding. For example, you can store the square root of the normalized value and then square it while decoding in order to increase precision on the low end of the range.

There's more...

It's a shame to leave the three components in the last render target unused. Among the things you can store in those components are:

- ▶ Per-pixel X and Y velocity for motion blur
- ▶ Ambient intensity mask
- ▶ Directional shadow value
- ▶ Material ID

GBuffer unpacking

Before any light related work can be accomplished, all the data stored in the GBuffer has to be unpacked back into a format that we can use with the lighting code presented in the forward lighting recipes.

You may have noticed one thing missing from our GBuffer and that is the world space position needed for lighting. Although we could store per-pixel world positions in our GBuffer, those positions are implicitly stored already in the depth buffer as non-linear depth. Together with the projection parameters, we are going to reconstruct the linear depth of each pixel (world distance from the camera) as part of the GBuffer unpacking process. The linear depth will later be used together with the inversed view matrix to reconstruct the world position of each pixel.

Getting ready

While unpacking the GBuffer, you always want to use point sampling when fetching data from the GBuffer texture views. You will need to set one point sampler and the four shader resource views for the pixel shader.

As mentioned in the introduction, we will also want to reconstruct the per-pixel linear depth. In addition to the depth buffer value, we will need some of the projection parameters used in the projection matrix to accomplish this task. Those additional projection values will be stored in a constant buffer that should be set for the pixel shader.

How to do it...

In order to make use of the data in the GBuffer we will need to sample from it. For that we are going to define the following textures:

```
Texture2D DepthTexture   : register( t0 );
Texture2D ColorSpecIntTexture: register( t1 );
Texture2D NormalTexture: register( t2 );
Texture2D SpecPowTexture: register( t3 );
```

You will need to set the shader resource views in the same order the texture register deceleration.

Once we unpack the GBuffer values, we are going to store them in a structure. This will make it easier to pass that data to the lighting functions. That structure is defined as follows:

```
struct SURFACE_DATA
{
    floatLinearDepth;
    float3 Color;
    float3 Normal;
    floatSpecInt;
    floatSpecPow;
};
```

Next you will need to call the main unpacking function which looks like this:

```
SURFACE_DATA UnpackGBuffer(int2 location)
{
    SURFACE_DATA Out;

    // Cast to 3 component for the load function
    int3 location3 = int3(location, 0);

    // Get the depth value and convert it to linear depth
```

```
        float depth = DepthTexture.Load(location3).x;
        Out.LinearDepth = ConvertDepthToLinear(depth);

        // Get the base color and specular intensity
        float4 baseColorSpecInt = ColorSpecIntTexture.Load(location3);
        Out.Color = baseColorSpecInt.xyz;
        Out.SpecInt = baseColorSpecInt.w;

        // Sample the normal, convert it to the full range and noramalize it
        Out.Normal = NormalTexture.Load(location3).xyz;
        Out.Normal = normalize(Out.Normal * 2.0 - 1.0);

        // Scale the specular power back to the original range
        floatSpecPowerNorm = SpecPowTexture. .Load(location3).x;
        Out.SpecPow = SpecPowerNorm.x + SpecPowerNorm * g_SpecPowerRange.y;

        return Out;
}
```

This function takes the pixel coordinates for the pixel we are unpacking. The function returns the filled surface data structure.

Converting the stored depth into linear depth requires two components from the projection matrix. However, we will need two additional perspective components and the inversed perspective matrix to reconstruct the world position. To avoid confusion in later recipes we are going to include all of those in the constant buffer that is defined as follows:

```
cbuffercbGBufferUnpack : register( b0 )
{
   float4 PerspectiveValues : packoffset( c0 );
   float4 x4ViewInv         : packoffset( c1 );
}
```

You will need to set the `PerspectiveValues` variable with the following data:

- XY components should be filled with one over the projection matrix diagonal values (For example for X, store `PerspectiveValues .x =1 / Pm[0][0]`)
- Z component is the projection matrix `Pm[3][2]` value
- W component is the negative projection matrix `Pm[2][2]` value

The code to convert the sampled depth into linear depth is as follows:

```
float ConvertDepthToLinear(float depth)
{
    float linearDepth = PerspectiveValues.z / (depth +
    PerspectiveValues.w);
    return linearDepth;
}
```

This function takes the sampled depth as input and returns the linear depth.

How it works...

As you may have expected, most of the code in the shader reads back the data from the GBuffer views. The specular power gets scaled back to its original range by using the inversed operation with the same values. The only part that requires further explanation is the conversion to linear depth.

If you recall from the forward lighting chapter, the perspective matrix has the following form:

$$
\begin{bmatrix} w & 0 & 0 & 0 \\ 0 & h & 0 & 0 \\ 0 & 0 & Q & 0 \\ 0 & 0 & -QZ_n & 0 \end{bmatrix}
\quad
\begin{aligned}
w &= \cot\left(\frac{fov_w}{2}\right) \\
h &= \cot\left(\frac{fov_h}{2}\right) \\
Q &= \frac{Z_f}{Z_f - Z_n}
\end{aligned}
$$

The value stored in the depth buffer originates from the dot product of the world space position with the third column. All we have to do to convert the depth buffer value back to linear depth, which is just reverse this in the same way we do with the specular power.

Directional light

Now that we know how to decode the GBuffer, we can use it to calculate directional lighting. Unlike forward rendering, where we rendered the scene in order to trigger the pixel shader and do the light calculations, the whole point using a deferred approach is to avoid rendering the scene while doing the lighting. In order to trigger the pixel shader we will be rendering a mesh with a volume that encapsulates the GBuffer pixels affected by the light source. Directional light affects all the pixels in the scene, so we can render a full screen quad to trigger the pixel shader for every pixel in the GBuffer.

Getting ready...

Back in the old days of Direct3D 9 you would need to prepare a vertex buffer to render a full screen quad. Fortunately, this is a thing of the past. With DirectX11 (as well as DirectX10) it is very easy to render a full screen quad with null buffers and sort out the values in the vertex shader. You will still need a constant buffer to store the same light information we used for the forward directional light.

In addition to the light information, we will need all the settings we used for unpacking the GBuffer including the shader resource view and the constant buffer. It is recommended you separate the GBuffer unpacking constants from the light related constants as the GBuffer constants are not going to change through the lighting stage.

One last thing you will have to take care of is to disable depth and nullify the depth buffer. The reason for doing so is that the full screen quad is not going to benefit from depth testing as it is supposed to cover the whole visible area (there is also a restriction about sampling and writing to the same depth buffer, but we will discuss this in the next recipe).

How to do it...

As with the forward directional light, we are going to use a buffer with the directional light information. You only need to set this buffer to the pixel shader. The buffer is defined as follows:

```
cbuffercbPerObject : register( b0 )
{
    float3 DirToLight      : packoffset( c0 );
    float4 DirLightColor   : packoffset( c1 );
}
```

We will be using the same material structure used for forward lighting that is defined as follows:

```
struct Material
{
    float3 normal;
    float4 diffuseColor;
    float specIntensity;
};
```

In order to render a full screen quad, you will need to set both index and vertex buffers to null. Set the topology to triangle strip and call `Draw` with four vertices starting from position zero.

Our vertex shader output structure is defined as follows:

```
struct VS_OUTPUT
{
    float4 Position    : SV_Position;
    float2 UV        : TEXCOORD0;
};
```

We will be using a constant array in the vertex shader to help selecting the vertex clip-space positions for each vertex:

```
staticconst float2 arrBasePos[4] = {
    float2(-1.0, 1.0),
    float2(1.0, 1.0),
    float2(-1.0, -1.0),
    float2(1.0, -1.0),
};
```

Finally, our vertex shader is defined as follows:

```
VS_OUTPUT DirLightVS(uintVertexID : SV_VertexID )
{
  VS_OUTPUT Output;

  Output.Position = float4(arrPos[VertexID].xy, 0.0, 1.0);
  Output.cpPos = Output.Position.xy;

  return Output;
}
```

This function takes the vertex index (in our case a value between 0 and 6) and outputs the vertex projected position and clip-space XY positions.

Now that we sorted out the full-screen quad rendering, it's time to move on to the pixel shader code. Using all the helper functions presented during the GBuffer unpacking recipe plus the new function that reconstructs the world position, the work done by the shader is:

▶ Unpack the GBuffer

▶ Fill the material structure from the unpacked values

▶ Reconstruct the world position

▶ Calculate the ambient and directional light values

If you recall, we will be using the `EyePosition` function to find the vector from the pixel to the camera, which is used for specular highlight calculation. Fortunately, this value is already present in the inversed view matrix. All we need to do in order to use it is to add this define:

```
#define EyePositionViewInv[3].xyz;
```

The code that handles all the preceding steps is as follows:

```
float4 DirLightPS( VS_OUTPUT In ) : SV_TARGET
{
    // Unpack the GBuffer
    SURFACE_DATA gbd = UnpackGBuffer(In.Position, 0));
```

```
// Convert the data into the material structure
Material mat;
mat.normal = gbd.Normal;
mat.diffuseColor.xyz = gbd.Color;
mat.diffuseColor.w = 1.0; // Fully opaque
mat.specPow = g_SpecPowerRange.x + g_SpecPowerRange.y * gbd.SpecPow;
mat.specIntensity = gbd.SpecIntensity;

// Reconstruct the world position
float3 position = CalcWorldPos(In.cpPos, gbd.LinearDepth);

// Calculate the ambient and directional light contributions
float4 finalColor;
finalColor.xyz = CalcAmbient(mat.normal, mat.diffuseColor.xyz);
finalColor.xyz += CalcDirectional(position, mat);
finalColor.w = 1.0; // Pixel is fully opaque

returnfinalColor;
}
```

This function takes the output from the vertex shader, which includes the pixel texture position and clip space position, and returns the pixels shaded color.

Finally, the function that reconstructs the world position is as follows:

```
float3 CalcWorldPos(float2 csPos, float linearDepth)
{
    float4 position;

    position.xy = csPos.xy * PerspectiveValues.xy * linearDepth;
    position.z = linearDepth;
    position.w = 1.0;

    return mul(position, ViewInv).xyz;
}
```

This function takes the clip-space position and the linear depth as inputs and returns the original world position the depth value originated from.

How it works...

Rendering the full screen quad exploits the fact that you can now issue a draw call without using buffers. This option was added to DirectX10 as part as the new **UAV** (**Unordered Access View**) buffer. In a situation where all the required input data to a vertex shader is coming from a UAV, this new feature prevents the need to bind a vertex buffer even though you don't need it.

Reconstructing the world position is pretty straightforward. Once we have the linear depth, we manually construct the view space position by inverting the projection effect on the clip-space position. Then, using the inversed view matrix we can transform the view-space position back to world space.

Point light

Compared to the directional light, point lights may only cover parts of the screen. In addition, it may be partially or fully occluded by the scene as viewed when generating the GBuffer. Fortunately, all we need to represent the light source volume is to render the front of a sphere located at the point light's center position and scale to its range.

Representing the light source as its volume makes sense when you think about the portion of the scene encapsulated by this volume. In the point lights case, the portion of the scene that gets lit is going to be inside this sphere volume. Rendering the volume will trigger the pixel shader for all pixels that are inside the volume, but could also trigger the pixel shader for pixels that are in front or behind the volume. Those pixels that are outside the volume will either get culled by the depth test or get fully attenuated so they won't affect the final image.

In the DirectX9 days you would normally use a pre-loaded mesh to represent the sphere. DirectX11 however, opens the possibility to render the volume by using the tessellation feature without any pre-loaded sphere mesh. After comparing the GPU performance cost for rendering the light volumes in both methods, I came to the conclusion that the HLSL generated volume is the right way to go.

Getting ready

Similar to forward lighting, we want to accumulate the combined lit output of all the light sources. Once the directional (with ambient) light is rendered, we will be using additive alpha blending to accumulate the light values from all the other light types. You will need to set the blend state before rendering any point lights and keep using it until all the lights are rendered.

Rendering the point lights volume will require a different rasterizer and depth stencil states than what we have used before. While rendering the point light volume we will be culling front faces (normally, we cull back faces) and do a Greater Than Equals depth comparison test (usually, we do Less Than Equals). This will help us avoid culling some or all of the point light volume faces when the camera is inside the volume.

To create the new depth stencil state, use the following descriptor:

Depth stencil descriptor parameter	Value
DepthEnable	TRUE
DepthWriteMask	D3D11_DEPTH_WRITE_MASK_ZERO
DepthFunc	D3D11_COMPARISON_GREATER_EQUAL

The reset of the parameters should be set to zero.

To create the rasterizer state, use the following descriptor:

Rasterizer descriptor parameter	Value
FillMode	D3D11_FILL_SOLID
CullMode	D3D11_CULL_BACK

The reset of the parameters should be set to zero.

We need to prepare a matrix that will transform the light volume from its local space to the projected space. To prepare that matrix, you will need to have the following information:

- Light position
- Light range
- Camera view matrix
- Camera projection matrix

To combine those values into a single matrix, first prepare a scale matrix with the light range as the scale factor. You will probably want to add a small value to the range to take into account the fact that the light volume is not perfectly round (more on that in the *How it works...* section). The range matrix has the following structure (**Lr** is the light range):

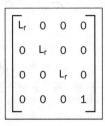

$$\begin{bmatrix} L_r & 0 & 0 & 0 \\ 0 & L_r & 0 & 0 \\ 0 & 0 & L_r & 0 \\ 0 & 0 & 0 & 1 \end{bmatrix}$$

Use the light position to prepare a translation matrix that has the following structure (**Lx**, **Ly**, and **Lz** are the lights position components):

$$
\begin{bmatrix}
1 & 0 & 0 & 0 \\
0 & 1 & 0 & 0 \\
0 & 0 & 1 & 0 \\
Lx & Ly & Lz & 1
\end{bmatrix}
$$

Remove the rotation component from the view matrix (the volume will always face the camera). This will give us only the translation portion that has the following matrix structure (**Tx**, **Ty**, and **Tz** are the translation components that should be kept):

$$
\begin{bmatrix}
1 & 0 & 0 & 0 \\
0 & 1 & 0 & 0 \\
0 & 0 & 1 & 0 \\
Tx & Ty & Tz & 1
\end{bmatrix}
$$

The projection matrix should not be modified. Finally, you have to multiply all those matrices. The multiplication order is:

LightRangeScale * LightPositionTranslation * ViewTranslation * Projection

This combined matrix will be sent to the Domain shader via a constant buffer.

As with the directional light, we will need all the GBuffer values and a constant buffer with the point light information. Note that if you don't reset the GBuffer views and constant buffer, all those values are still going to be set, so you need to do is only set the constant buffer with the light information.

Once all the values are ready and the shaders are set, all you need to do is issue a call to draw with a single vertex (as with the directional light, we won't be using any vertex or index buffers).

How to do it...

Before we dive into the HLSL code we have to set a different primitive topology. As we will be using the tessellation feature we will be using the following topology: **D3D11_PRIMITIVE_ TOPOLOGY_1_CONTROL_POINT_PATCHLIST**. In addition to the topology, set the depth stencil and rasterizer states we created in the *Getting ready...* section.

We will start with the light volume generation by calling `Draw` with 2 and 0 as inputs. Unlike all the previous recipes, we will be doing most of the heavy lifting in the Domain shader. The vertex shader code is as follows:

```
float4 PointLightVS() : SV_Position
{
    return float4(0.0, 0.0, 0.0, 1.0);
}
```

This function takes no input and outputs a single position. We won't be using this position, but outputting a position is a minimum requirement for a vertex shader so we have to output something anyway.

Next we can define the constant hull shader. Don't feel intimidated if you are not familiar with hull shaders. You can find an explanation to how they work in the *How it works...* section of this recipe. This shader outputs the following structure which contains the tessellation factors for the patch edges and inside:

```
struct HS_CONSTANT_DATA_OUTPUT
{
    float Edges[4] : SV_TessFactor;
    float Inside[2] : SV_InsideTessFactor;
};
```

The hull constant shader looks like this:

```
HS_CONSTANT_DATA_OUTPUT PointLightConstantHS()
{
    HS_CONSTANT_DATA_OUTPUT Output;

    floattessFactor = 18.0;
    Output.Edges[0] = Output.Edges[1] = Output.Edges[2] =
    Output.Edges[3] = tessFactor;
    Output.Inside[0] = Output.Inside[1] = tessFactor;

    return Output;
}
```

This function takes no input and outputs a structure with the tessellation values we want to apply to the patch. This implementation of the hull shader is very simple, so we won't be using any of the inputs. To simplify things we will be using a constant value that provides a decent amount of smoothness to the half sphere.

Similar to the vertex shader, our hull shader will need to output a position, only this time we will use it to switch between the two sides of the sphere. We will be using the following constant array to help us switch between the two sides:

```
static const float3 HemilDir[2] = {
    float3(1.0, 1.0,1.0),
    float3(-1.0, 1.0, -1.0)
};
```

The code for the hull shader is as follows:

```
[domain("quad")]
[partitioning("integer")]
[outputtopology("triangle_ccw")]
[outputcontrolpoints(4)]
[patchconstantfunc("PointLightConstantHS")]
float3 PointLightHS(uint PatchID : SV_PrimitiveID) : POSITION
{
    HS_OUTPUT Output;

    Output.HemiDir = HemilDir[PatchID];

    return Output;
}
```

Basically, our hull shader will generate a patch with four control points that will get tessellated based on the value returned from the constant hull shader.

Once the tessellator is done, we will need a Domain shader to find the final position of each one of the generated quads. As always, the positions generated by the Domain shader have to be transformed into projected space. For that we will be using the matrix we built in the preceding section, which will be stored in the following constant buffer:

```
cbuffer cbPointLightDomain : register( b0 )
{
    float4x4 LightProjection : packoffset( c0 );
}
```

We would also need a structure for the Domain shader return values which is defined as follows:

```
struct DS_OUTPUT
{
    float4 Position    : SV_POSITION;
    float2 cpPos       : TEXCOORD0;
};
```

The Domain shader code is as follows:

```
[domain("quad")]
DS_OUTPUT PointLightDS( HS_CONSTANT_DATA_OUTPUT input, float2 UV : SV_
DomainLocation, constOutputPatch<HS_OUTPUT, 4> quad)
{
    // Transform the UV's into clip-space
    float2 posClipSpace = UV.xy * 2.0 - 1.0;

    // Find the absulate maximum distance from the center
    float2 posClipSpaceAbs = abs(posClipSpace.xy);
    float maxLen = max(posClipSpaceAbs.x, posClipSpaceAbs.y);

    // Generate the final position in clip-space
    float3 normDir = normalize(float3(posClipSpace.xy, maxLen - 1.0) *
    quad[0].HemiDir);
    float4 posLS = float4(normDir.xyz, 1.0);

    // Transform all the way to projected space and generate the UV
    coordinates
    DS_OUTPUT Output;
    Output.Position = mul(posLS, LightProjection );

    // Store the clip space position
    Output.cpPos = Output.Position.xy / Output.Position.w;

    return Output;
}
```

This function takes the UV coordinates generated by the tessellator as input and returns the projected and clip space positions of the vertex.

So far we have covered all the code needed for generating the light volume and transforming it to projected space. Now it's time to do the light calculations in the pixel shader. As with the forward point light, we will need a constant buffer with the lights information:

```
cbuffercbPointLightPixel : register( b1 )
{
    float3 PointLightPos    : packoffset( c0 );
    float  PointLightRange  : packoffset( c0.w );
    float3 PointColor       : packoffset( c1 );
    float  PointIntensity   : packoffset( c1.w );
}
```

Finally, the code for the actual pixel shader:

```
float4PointLightPS( DS_OUTPUT In ) : SV_TARGET
{
    // Unpack the GBuffer
    SURFACE_DATA gbd = UnpackGBuffer(In.Position.xy);

    // Convert the data into the material structure
    Material mat;
    mat.normal = gbd.Normal;
    mat.diffuseColor.xyz = gbd.Color;
    mat.diffuseColor.w = 1.0; // Fully opaque
    mat.specPow = g_SpecPowerRange.x + g_SpecPowerRange.y * gbd.SpecPow;
    mat.specIntensity = gbd.SpecIntensity;

    // Reconstruct the world position
    float3 position = CalcWorldPos(In.cpPos, gbd.LinearDepth);

    // Calculate the light contribution
    float4 finalColor;
    finalColor.xyz = CalcPoint(position, mat);
    finalColor.w = 1.0;

    return finalColor;
}
```

This function takes the clip space position and texture coordinates as input and outputs the lit pixel value. As you can see, this is almost the exact copy of the directional light, the only thing that changed is the call to the `CalcPoint` function. The `CalcPoint` function is the exact copy of the function we used for forward lighting.

How it works...

If you are not familiar with the DirectX11 tessellation feature, you should first look at the following diagram showing the various stages of a DirectX11 enabled GPU:

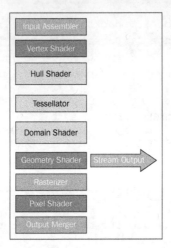

The new stages marked in green handle all the tessellation work for us. Both **Hull Shader** and **Domain Shader** are programmable, but the **Tessellator** isn't. As you recall, our goal was to build a half-sphere mesh facing the camera. We started off from a single vertex generated from the call to `Draw(1, 0)`. Our vertex shader just passed the vertex position to the Hull shade and there is where the magic happens.

A **Hull Shader** takes care of two tasks. First, it tells the **Tessellator** how to subdivide the output shape (in our case a quad). Second, it passes information about the control points, such as positions and normal, to the **Domain Shader**. In this recipe, all we needed from the **Hull Shader** is to tell the **Tessellator** to split a quad into a grid.

Once the **Tessellator** comes up with the division, it will trigger our **Domain Shader** once for each vertex in the newly generated grid. Among the input to the **Domain Shader**, we receive the UV which is a two component normalized value. Each vertex will get a unique UV value which represents its position on the grid. Our **Domain Shader** uses this value to wrap the grid as a half-sphere.

In order to wrap the grid, we need to calculate the 3D half-sphere positions from a 2D grid position. For that we can take advantage of the fact that normalizing a vector converts it to a unit length. We convert the UV into a space similar to clip-space (range is converted from [0, 1] to [-1, 1]). Now we need to come up with a third component which will be our positions' Z value. As the value of Z increases towards the center of the grid and becomes zero at its edges, we can use the maximum absolute value between X and Y. Finally, we can normalize this vector and get the vertices position on the half-sphere. Once we get this 3D position, the matrix will scale the half sphere to the light range size and position it at the lights location.

You can visualize this transformation from grid to half-sphere by looking at the following image showing the grid on the left and the half-sphere on the right:

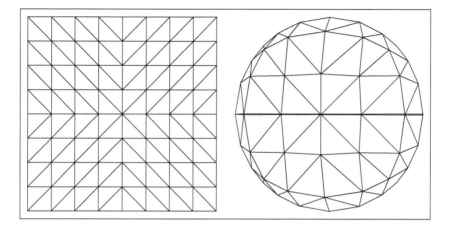

When setting the scale of the matrix, you should use a value slightly higher than the lights range. As you can see from the preceding image, the half-sphere is not perfect and contains straight edges. If we set the scale to exactly the lights range, the vertices will be positioned at the edge of the lights volume, but the edges connecting those vertices will cut into the lights volume.

Capsule light

Capsule light volume is made of a sphere divided into two and extruded along the capsules segment to form the cylindrical part of the capsule. As with the point light volume, we will use the tessellator to generate the capsule volume, only this time we will expend two control points instead of one. Each one of these points will eventually get turned into a half sphere attached to half of the capsules cylindrical part. The two capsule halves will connect at the center of the capsule.

Getting ready...

Our capsule will be scaled by two values, a range and a segment length. Those are the exact same values we used for the forward capsule light. Unfortunately, we can't use a matrix to scale the capsule because we need to scale the segment and the range separately. Therefore, we will set the capsule lights range, segment length, and a transformation to projected space to the Domain shader. The matrix will combine the capsule rotation and translation with the view and projection matrix.

In addition to the matrix and the usual GBuffer related shader constants, we will need the same light parameters we used in the forward capsule light.

How to do it...

To start the rendering the capsule light, we will call `Draw` with two vertices from offset zero. This will invoke the vertex shader twice. As with the point and spot light, we don't care about the position the vertex shader will output, so we can use the same vertex shader we used for the other two light types.

In order to expend each vertex into a different half of the capsule, our Hull shader will need to output a different direction for each input control point. As we are forced to output a position, we will be using this output to specify the capsule direction. The new output structure for the Hull shader is as follows:

```
struct HS_OUTPUT
{
    float4 CapsuleDir : POSITION;
};
```

We will store the two direction vectors in the following shader constant:

```
staticconst float4 CapsuelDir[2] = {
    float4(1.0, 1.0, 1.0, 1.0),
    float4(-1.0, 1.0, -1.0, 1.0)
};
```

Out Hull shader will select the direction based on the primitive index, which is just the index of the vertex:

```
[domain("quad")]
[partitioning("integer")]
[outputtopology("triangle_ccw")]
[outputcontrolpoints(4)]
[patchconstantfunc("CapsuleLightConstantHS")]
HS_OUTPUT CapsuleLightHS(uintPatchID : SV_PrimitiveID)
{
    HS_OUTPUT Output;

    Output.CapsuleDir = CapsuelDir[PatchID];

    return Output;
}
```

This function takes the primitive ID as input and outputs the capsule direction.

Our constant Hull shader is going to use a tessellation factor of 12 for both edges and inside values.

We will be using three constants to split each quad into the half sphere and half cylinder portion in the Domain shader. We will be using 30 percent of the capsule length on the z axis for the half cylinder and the rest for the half sphere. Those constants with the Domain shader code look like this:

```
#define CylinderPortion 0.3
#define SpherePortion   (1.0 - CylinderPortion)
#define ExpendAmount    (1.0 + CylinderPortion)
```

Our Domain shader constant buffer that contains the capsule transformation to projected space along with half the length of the capsule segment and the range is defined as follows:

```
cbuffercbSpotLightDomain : register( b0 )
{
  float4x4 LightProjection : packoffset( c0 );
  float HalfSegmentLen      : packoffset( c4 );
  float CapsuleRange        : packoffset( c4.y );
}
```

The Domain shader that uses these values to generate the capsule half:

```
[domain("quad")]
DS_OUTPUT CapsuleLightDS( HS_CONSTANT_DATA_OUTPUT input, float2 UV :
SV_DomainLocation, constOutputPatch<HS_OUTPUT, 4> quad)
{
    // Transform the UV's into clip-space
    float2 posClipSpace = UV.xy * 2.0 - 1.0;

    // Find the vertex offsets based on the UV
    float2 posClipSpaceAbs = abs(posClipSpace.xy);
    float maxLen = max(posClipSpaceAbs.x, posClipSpaceAbs.y);
    float2 posClipSpaceNoCylAbs = saturate(posClipSpaceAbs *
    ExpendAmount);
    float2 posClipSpaceNoCyl = sign(posClipSpace.xy) *
    posClipSpaceNoCylAbs;
    float maxLenNoCapsule = max(posClipSpaceNoCylAbs.x,
    posClipSpaceNoCylAbs.y);

    // Generate the final position in clip-space
    float3 normDir = normalize(float3(posClipSpaceNoCyl.xy,
    maxLenNoCapsule - 1.0)) * CapsuleRange;
    float cylinderOffsetZ = saturate(maxLen - min(maxLenNoCapsule,
    SpherePortion)) / CylinderPortion;
    float4 posLS = float4(normDir.xy, normDir.z + cylinderOffsetZ *
    HalfSegmentLen- HalfSegmentLen, 1.0);

    // Move the vertex to the selected capsule side
```

```
            posLS *= quad[0].CapsuleDir;

            // Transform all the way to projected space and generate the UV
            coordinates
            DS_OUTPUT Output;
            Output.Position = mul(posLS, LightProjection );
            Output.UV = Output.Position.xy / Output.Position.w;
            Output.UV = Output.UV * float2(0.5, -0.5) + 0.5;

            return Output;
        }
```

This function takes the tessellator UV values and the capsule direction selected by the Hull shader as input and returns the projected capsule vertex with its matching UV coordinates.

As with the other deferred light types, we are going to use the same constant buffer and function to do the actual lighting calculation in the pixel shader. Our pixel shader looks similar to the other light types other than the call to `CalcCapsule`:

```
    float4 CapsuleLightPS( DS_OUTPUT In ) : SV_TARGET
    {
        // Unpack the GBuffer
        SURFACE_DATA gbd = UnpackGBuffer(In.UV);

        // Convert the data into the material structure
        Material mat;
        mat.normal = gbd.Normal;
        mat.diffuseColor.xyz = gbd.Color;
        mat.diffuseColor.w = 1.0; // Fully opaque
        mat.specPow = g_SpecPowerRange.x + g_SpecPowerRange.y * gbd.SpecPow;
        mat.specIntensity = gbd.SpecIntensity;

        // Reconstruct the world position
        float2 cpPos = In.UV.xy * float2(2.0, -2.0) - float2(1.0, -1.0);
        float3 position = CalcWorldPos(cpPos, gbd.LinearDepth);

        // Calculate the light contribution
        float4 finalColor;
        finalColor.xyz = CalcCapsule(position, mat);
        finalColor.w = 1.0;

        return finalColor;
    }
```

This function takes the Domain shader output and returns the lit pixel color.

How it works...

Generating the capsule volume is not different from the technique we used for the point light volume. We basically expend a single point to a tessellated grid. The grid points are positioned in a half sphere formation, but unlike the point light volume, we force some of the points to overlap at the outline of the half sphere open end. We then proceed to scale all the points to the lights range. Finally, we move all the extra overlapping vertices along the z axis to get the half cylinder shape. By repeating the same process and inverting it on the z axis, we get the capsule shape. The following image shows the resulting capsule rendered in wireframe:

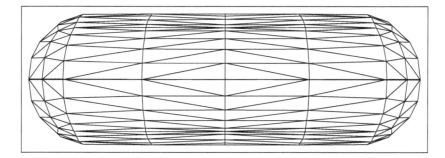

After the vertices get positioned and scaled by the segment and range values, we apply the transformation matrix that will rotate and translate the capsule and transform it to projected space.

Spot light

Spot light volume is represented by a cone with a rounded base. We can use same technique used for generating the capsule light half capsule volume and shape the result as the spot lights cone. As this requires only half of the capsule, we will only need to draw a single vertex.

Getting ready

As with the point and capsule lights, we will need all the GBuffer related values along with the pixel shader constants we used for the light properties in the forward spot light recipe.

Unlike the capsule light volume, the spot light volume can be scaled by a matrix, so we won't need to pass any scaling parameters to the Domain shader. The following illustration shows how the pretransform cone will look like:

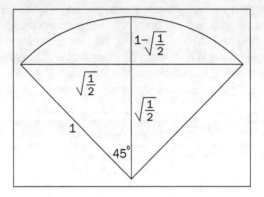

We can use the following formula to calculate the matrix X and Y scale value:

$$ScaleXY = \frac{LightRange * tana}{\sqrt{\frac{1}{2}}}$$

Where α is the cone's outer angle **LightRange** is the spot light's range. For the Z component we just use the light range value.

The light cone tip will be at coordinate origin, so the rotation and translation parts of the Domain shader matrix can be calculated in the same way we did for the capsule light, only use the spot light direction instead of the capsule segment direction. Once the scale, rotation and translation are ready, we can combine them with the view and projection matrix to get the final transformation matrix in the following order:

FinalMatrix = Scale * Rotate * Translate * View * Projection

How to do it...

As with the point light, we will issue the `Draw` call for each spot light by calling `Draw(1, 0)`. We will be using the same Vertex shader, Hull shader, and Constant Hull shader that we used for point light volume generation. Similar to the capsule light volume, we will be using a tessellation value of 12.

Our Domain shader will output the same output structure we used for both point and spot lights and will also use the capsule portion constant defined in the capsule light recipe. In addition, it will use the following two constant values:

```
#define CylinderPortion 0.2
#define ExpendAmount    (1.0 + CylinderPortion)
```

The code for the Domain shader is as follows:

```
[domain("quad")]
DS_OUTPUT SpotLightDS( HS_CONSTANT_DATA_OUTPUT input, float2 UV : SV_
DomainLocation, const OutputPatch<HS_OUTPUT, 4> quad)
{
    // Transform the UV's into clip-space
    float2 posClipSpace = UV.xy * float2(2.0, -2.0) +
    float2(-1.0, 1.0);

    // Find the vertex offsets based on the UV
    float2 posClipSpaceAbs = abs(posClipSpace.xy);
    float maxLen = max(posClipSpaceAbs.x, posClipSpaceAbs.y);

    // Force the cone vertices to the mesh edge
    float2 posClipSpaceNoCylAbs = saturate(posClipSpaceAbs *
    ExpendAmount);
    float maxLenNoCapsule = max(posClipSpaceNoCylAbs.x,
    posClipSpaceNoCylAbs.y);
    float2 posClipSpaceNoCyl = sign(posClipSpace.xy) *
    posClipSpaceNoCylAbs;

    // Convert the positions to half sphere with the cone vertices on
    the edge
    float3 halfSpherePos = normalize(float3(posClipSpaceNoCyl.xy,
    1.0 - maxLenNoCapsule));

    // Scale the sphere to the size of the cones rounded base
    halfSpherePos = normalize(float3(halfSpherePos.xy * SinAngle,
    CosAngle));

    // Find the offsets for the cone vertices (0 for cone base)
    float cylinderOffsetZ = saturate((maxLen * ExpendAmount - 1.0) /
    CylinderPortion);

    // Offset the cone vertices to their final position
    float4 posLS = float4(halfSpherePos.xy * (1.0 - cylinderOffsetZ),
    halfSpherePos.z - cylinderOffsetZ * CosAngle, 1.0);

    // Transform all the way to projected space and generate the UV
    coordinates
    DS_OUTPUT Output;
```

```
Output.Position = mul( posLS, LightProjection );
Output.cpPos = Output.Position.xy / Output.Position.w;

    return Output;
}
```

This function takes the Hull shader outputs with the tessellator UV coordinates and returns the spot volume vertex position in projected space and the vertex UV coordinates.

For the pixel shader, we will be using the exact code we did for the point light, only this time we will be calling the `CalcSpot` function (instead of the call to the `CalcPoint` function). This will give us the following code:

```
float4 SpotLightPS( DS_OUTPUT In ) : SV_TARGET
{
    // Unpack the GBuffer
    float2 UV = In.UV; // This is not working so well...
    SURFACE_DATA gbd = UnpackGBuffer(UV);

    // Convert the data into the material structure
    Material mat;
    mat.normal = gbd.Normal;
    mat.diffuseColor.xyz = gbd.Color;
    mat.diffuseColor.w = 1.0; // Fully opaque
    mat.specPow = g_SpecPowerRange.x + g_SpecPowerRange.y * gbd.SpecPow;
    mat.specIntensity = gbd.SpecIntensity;

    // Reconstruct the world position
    float2 cpPos = UV.xy * float2(2.0, -2.0) - float2(1.0, -1.0);
    float3 position = CalcWorldPos(cpPos, gbd.LinearDepth);

    // Calculate the light contribution
    float4 finalColor;
    finalColor.xyz = CalcSpot(position, mat);
    finalColor.w = 1.0;

    return finalColor;
}
```

This function takes the Domain shader output and returns the light pixel color.

How it works...

Most of the Domain shaders presented in this recipe are very similar to the half capsule generation code from the capsule light recipe. However, the spot light cone has two things that are different: The cylinder portion should be a cone and the rounded base of the cone is a smaller portion of a sphere.

If you look at the illustration showing the side view of the cone, you may notice that the rounded base of the cone has a constant radius of one from the cone tip. By first creating a sphere, scaling the Z components and normalizing again our Domain shader reshapes the half sphere positions to have a distance of one from the tip of the cone, which gives us the rounded base we need. The scale used is the length from the flat base of the cone to the center of the rounded base.

With the rounded base ready, we proceed to extrude the duplicate vertices into a cylinder like we did with the half capsule. By linearly scaling the XY components inwards along the z axis, the cylinder turns into the cone.

The following image shows how the capsule volume looks like from a side view when rendered in wireframe mode:

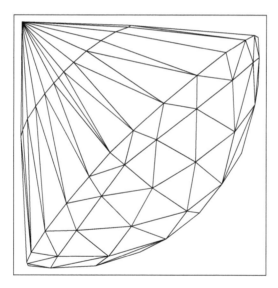

3

Shadow Mapping

In this chapter we will cover various shadow mapping techniques:

- ▶ Spot light PCF shadows
- ▶ Point light PCF shadows
- ▶ Cascaded shadow maps
- ▶ PCF with varying penumbra size
- ▶ Visualizing shadow maps

Introduction

Over the previous two chapters, we covered various ways to light a scene. The Blinn-Phong light model we used in all those recipes provides a good approximation to how light interacts with the rendered scene's surfaces. However, one major shortcoming of this light model is that it doesn't take into account occlusion by the various scene elements. In other words, it lacks shadows.

In real life, when a light ray hits an opaque surface, it gets reflected or absorbed and does not continue in the same direction. The Blinn-Phong model does not take this into consideration, so every surface inside the light's range will get illuminated even when it is occluded from the light source by other surfaces. Fortunately, shadows can be added to the Blinn-Phong light model as an attenuation factor. This chapter is going to focus on two shadow techniques that can be used for the various light sources featured in the two previous chapters.

In order to better understand how real-world shadows work, let's look at the following illustrated scenario showing a sphere-shaped light source, a shadow caster, and a shadow receiver:

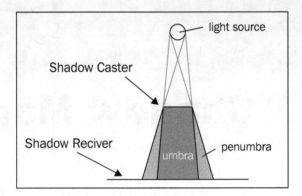

As you can see in the illustration, the surface marked as the shadow caster blocks some of the light traveling towards the surface marked as the receiver that introduces shadows. Since the light source has a sphere-shaped volume, the shadow is split into two areas:

- An area fully occluded from the light source called umbra
- An area partially occluded from the light called penumbra

Inside the penumbra, the closer a point is to the fully lit area, the more light it will receive. This makes the transition between the fully lit area and the umbra gradual.

Before GPUs commonly supported render targets, the main stream's shadow-calculation technique for games was called **Stencil Shadows**. In this technique, before calculating the light contribution of a light source, the stencil buffer was used to store a per-pixel boolean `is in shadow` value. During the lighting calculation, those pixels that where marked as `in shadow` were just skipped and the light didn't affect them. This process had to be repeated for every light source in the scene. Unfortunately, this Boolean `is in shadow` value means a pixel is either lit or is in shadow, so this method lacks any form of penumbra.

[Shadow algorithms that lack penumbra are referred to as "hard shadow" due to the sharp transition between lit and shadowed area. Due to their unnatural look they are not used anymore.]

Once render targets became commonly supported by GPUs, the trend switched away from the stencil buffer solution and render targets were used instead. Render target and programmable shaders opened new possibilities for shadow calculation and a lot of research was invested towards improving shadow map-based techniques. Hard shadows were quickly replaced with techniques that approximate the penumbra in various ways of success.

As you may have guessed, shadowing algorithms that approximate penumbras are referred to as "soft shadows". All the recipes in this chapter will focus on soft shadow implementations.

Render target shadow mapping techniques are generally referred to as shadow mapping. This chapter will cover ways to integrate shadow mapping with the various light sources featured in the previous chapters. It is important to understand that unlike deferred lights which has very low performance cost, shadow mapping is both performance and memory intensive. In fact, in scenes that contain a very large amount of light sources, it is usually impossible to support shadow casting for each and every light source. Due to the significant additional per-light cost introduced by adding shadow support, all the recipes in this chapter will provide a mechanism to toggle the shadow calculation on and off as needed.

Spot light PCF shadows

Out of all the different light sources we covered, adding shadow-casting support to spot lights is the easiest to accomplish. We will be using a technique based on the way projected textures are handled (covered in *Chapter 1, Forward Lighting*). This shadow-casting technique is called **percentage-closer filtering** (**PCF**).

The following screenshot features a side-by-side view of the same spot light with no shadows on the left-hand side and with shadows on the right-hand side:

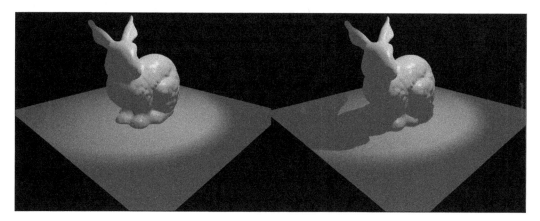

One side benefit of shadows is that they help clarify the relative height between different scene elements. In the left image of the screenshot, it's hard to tell if the bunny is floating in the air or standing on the surface. The shadows on the right image help us determine that the bunny is standing on the surface.

Getting ready

Similar to the way we handled projected textures, we will need to generate a transformation from world space to the light's projected view space for each spotlight-casting shadow. This transformation is generated from a combined view and projection transformations similar to the work of scene-view projection matrices, only this time those transformations are going to be based on the light configuration instead of the cameras.

In order to generate the light's view matrix, you will need a normalized random direction vector that is not parallel to the light's forward direction. The following illustration shows the various components of the required matrix:

$$
\begin{bmatrix}
R_x & R_y & R_z & 0 \\
U_x & U_y & U_z & 0 \\
D_x & D_y & D_z & 0 \\
P_x & P_y & P_z & 1
\end{bmatrix}
$$

We can now fill the previous matrix by following these descriptions:

- R: This is the cross product between the light direction and the random normal.
- U: This is the cross product between the light direction and R.
- D: This is the light direction in world space.
- P: This can be calculated by transforming the light's world space by the previous matrix with all P components set to 0. Once you get the transformed results, set them into P instead of the 0 value.

Use the following illustration to generate the projection matrix:

$$
\begin{bmatrix}
s & 0 & 0 & 0 \\
0 & s & 0 & 0 \\
0 & 0 & Q & 1 \\
0 & 0 & -QZ_n & 0
\end{bmatrix}
\qquad
\begin{aligned}
S &= \cot\left(\frac{fov}{2}\right) \\[2mm]
Q &= \frac{Z_f}{Z_f - Z_n}
\end{aligned}
$$

Use the following values to fill the projection matrix:

- `fov`: This is the spot light's field of view
- `zn`: This is the distance to the near plane (see explanation in the *How it works...* section)
- `zf`: This is the distance to the far plane

Please refer to the *Projected texture – spot light* recipe of *Chapter 1, Forward Lighting*, if you need to go over the full explanation of how those matrices work.

We will be rendering the scene using those two matrices into a depth render target called the shadow map. This shadow map will be used during the lighting calculations to find the shadow's attenuation value for each pixel getting lit. Depending on the range of the light and the complexity of the scene, you may either use a single channel texture of 16 or 32 bit for storage. You will need to prepare the following configuration:

	16-Bit	32-Bit
Texture	`DXGI_FORMAT_R16_TYPELESS`	`DXGI_FORMAT_R32_TYPELESS`
Depth Stencil View	`DXGI_FORMAT_D16_UNORM`	`DXGI_FORMAT_D32_UNORM`
Shader Resource	`DXGI_FORMAT_R16_UNORM`	`DXGI_FORMAT_R32_UNORM`

During the lighting calculations, we will be sampling the shadow map to come up with the shadow attenuations value. Before DirectX 10 was released, the PCF attenuation value had to be calculated manually in the pixel shader. From DirectX10 onwards, due to the popularity of PCF shadow mapping gained, a hardware implementation for PCF was introduced, which provides superior performance compared to previous implementations. You will need to define a sampler that uses the following description parameters:

Parameter	Value
Filter	`D3D11_FILTER_COMPARISON_MIN_MAG_MIP_LINEAR`
Address U/V/W	`D3D11_TEXTURE_ADDRESS_BORDER`
ComparisonFunc	`D3D11_COMPARISON_LESS`
BorderColor 0-3	`1.0`

How to do it...

We will start off by rendering the scene into the depth view. For performance reasons it is recommended that you only render the portion of the scene, which overlaps with the light source that affects volume (this is the same type of optimization we used for forward lights). Clear the depth view to 1.0, which is the nonlinear depth value for the far plane. You will have to set null views for the render targets along with the depth view, as we won't be outputting any colors.

Last thing to prepare is the rasterizer state with the depth bias values. You will need to set both, depth bias and slope scale depth bias, in the descriptor. This rasterizer state should only be used while rendering into the shadow map, so you should restore the previous state once this step is completed. See the full explanation on why we need the bias values and how to tune them in the *How it works...* section.

Once all the settings are ready and the list of visible meshes is prepared, we can start rendering them into the depth render target. Instead of the scene's view-projection transformation, we will be using the light's view-projection transformation we prepared in the *Getting Ready* section. For each mesh, you may use the same shaders that you would normally use to output the vertex positions, but it is recommended you prepare a separate set of shaders that only output the position without any other values normally passed to the pixel shader. Since we are only writing depths, you will not need a pixel shader unless you use alpha testing, so just clear the previous pixel shader by binding a null one.

The GPU will perform much faster when only outputting depth values in what's called "Double Speed Z" mode. This mode is turned on when no pixel shader is used and a depth map is bound. You can take advantage of this mode for shadow mapping and for priming the scene's depth.

With the shadow map filled with the shadow caster's depth values, we can now use it for shadow calculations for both, forward and deferred lighting. The following code changes should work for both types of light calculations. Note that it is better for performance to prepare all the shadow maps before starting the lighting calculations rather than preparing the shadow map and doing the light calculations for each light source separately.

As mentioned in the introduction, we want to support both, shadow casting and nonshadow casting light sources. In order to support both cases, separate the pixel shader entry point (the pixel shader function we pass to the compiler) for each case and merge the functionality under a common function as shown in the following code:

```
float4 SpotLightPS( DS_OUTPUT In ) : SV_TARGET
{
  return SpotLightCommonPS(In, false);
}

float4 SpotLightShadowPS( DS_OUTPUT In ) : SV_TARGET
{
  return SpotLightCommonPS(In, true);
}
```

Those two new entry points should be compiled into two pixel shaders, where the Boolean value passed `SpotLightCommonPS` and switch the shadow casting on or off. For the lights that are not casting shadows, we will use `SpotLightPS`, and for lights that perform cast shadows, we will use `SpotLightShadowPS` as the entry point.

With the entry points passing the Boolean value to the common function `SpotLightCommonPS`, we will need to get this function to pass the Boolean value into the spot calculation function as shown in the following code:

```
float4 SpotLightCommonPS ( DS_OUTPUT In, boolbUseShadow )
{
   ...
finalColor.xyz = CalcSpot (position, mat, bUseShadow);
   ...
}
```

Before we can go over the changes to the function `CalcSpot`, we need to add the shadow map texture and sampler definitions to the pixel shader, so we can use it in the PCF calculations. When using deferred lighting, you can bind the texture to slot 3 (slots 0 to 2 contain GBuffer) and the sampler to slot 1 (slot 0 is used for unpacking GBuffer). The shadow map and new sampler are defined as follows:

```
Texture2D<float> SpotShadowMapTexture     : register( t4 );
SamplerComparisonState PCFSampler  : register( s1 );
```

In order to use the shadow map to calculate the shadow attenuation, we will add the following function to the pixel shader, which calculates the PCF value for a given UV and depth value:

```
float SpotShadowPCF ( float3 position )
{
  // Transform the world position to shadow projected space
  float4 posShadowMap = mul (float4 (position, 1.0), ToShadowmap);

  // Transform the position to shadow clip space
  float3 UVD = posShadowMap.xyz / posShadowMap.w;

  // Convert to shadow map UV values
  UVD.xy = 0.5 * UVD.xy + 0.5;
  UVD.y = 1.0 - UVD.y;

  // Compute the hardware PCF value
  return SpotShadowMapTexture.SampleCmpLevelZero (PCFSampler, UVD.xy,
  UVD.z);
}
```

The last change we need to make is to the constant buffer used by the pixel shader. We will add the matrix we prepared in the *Getting Ready* section to the end of the constant buffer as follows:

```
cbuffer cbSpotLightPixel : register( b1 )
{
    float3 SpotLightPos       : packoffset( c0 );
    float SpotLightRangeRcp    : packoffset( c0.w );
    float3 SpotDirToLight     : packoffset( c1 );
    float SpotCosOuterCone     : packoffset( c1.w );
    float3 SpotColor          : packoffset( c2 );
    float SpotCosConeAttRange   : packoffset( c2.w );
    float4x4 ToShadowmap      : packoffset( c3 );
}
```

For lights that don't cast shadows, this matrix can be set to identity as it would not be used.

Now we can finally look at the changes to `CalcSpot`, which are covered by the following code:

```
float3 CalcSpot(float3 position, Material material, bool bUseShadow)
{
    float3 ToLight = SpotLightPos - position;
    float3 ToEye = EyePosition - position;
    float DistToLight = length(ToLight);

    // Phong diffuse
    ToLight /= DistToLight; // Normalize
    float NDotL = saturate(dot(ToLight, material.normal));
    float3 finalColor = material.diffuseColor.rgb * NDotL;

    // Blinn specular
    ToEye = normalize(ToEye);
    float3 HalfWay = normalize(ToEye + ToLight);
    float NDotH = saturate(dot(HalfWay, material.normal));
    finalColor += pow(NDotH, material.specPow) * material.specIntensity;

    // Cone attenuation
    float cosAng = dot(SpotDirToLight, ToLight);
    float conAtt = saturate((cosAng - SpotCosOuterCone) /
    SpotCosConeAttRange);
```

```
conAtt *= conAtt;

    float shadowAtt;
if(bUseShadow)
{
  // Find the shadow attenuation for the pixels world position
  shadowAtt = SpotShadowPCF(position);
}
else
{
  // No shadow attenuation
  shadowAtt = 1.0;
}

// Attenuation
float DistToLightNorm = 1.0 - saturate(DistToLight *
SpotLightRangeRcp);
float Attn = DistToLightNorm * DistToLightNorm;
finalColor *= SpotColor.rgb * Attn * conAtt * shadowAtt;

// Return the final color
return finalColor;
};
}
```

As you can see, the input Boolean `bUseShadow` is used in a branch that skips all the shadow-related code when equals to `false`. Otherwise, the new function `ShadowPCF` gets called and the returned value is combined with the attenuation, which adds the shadows.

When the HLSL compiler encounters a branch with a fixed value in the condition (this means that the value is hardcoded in the shader), it removes the branch and compiles only the code inside the path matching the condition result. This optimization means we won't pay any runtime performance price. Note that this optimization won't work for values passed to the shader as constants, as the shader compiler can't predict the value that will be passed in during runtime.

How it works...

Before we explain how the featured code changes work, it's important to understand how shadow mapping and PCF work. The PCF calculation used in this recipe works in two stages: preparing the shadow map before the lighting calculations and using it during the lighting calculations. The following illustration features a common scenario which would help explain those two stages:

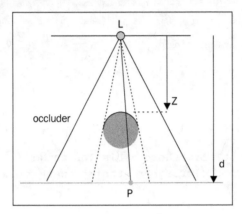

During the first stage, the depths of the scene portion affected by the light are rendered into the shadow map from the light's point of view. In our example, the light source L is a spot light, so the affected scene area is encapsulated by the black cone. With depth test enabled, the values written to the shadow map are those colored gold.

Once the shadow map is ready, the lighting calculations use it to determine if a pixel being rendered is in shadow or not. For a pixel located at point P with a depth value of d, the shadow map will be sampled and return the value z. Since z is closer than d to the light, the pixel is in shadow. In fact, any pixel with a position-colored green will end up in shadow, while any pixel with a position-colored gold will end up in light. If you remember how real-life shadows work, you may have noticed that this calculation is missing the penumbra portion of the shadow. The penumbra is not fully lit nor is it fully in shadow, so our Boolean value won't be enough to describe it. In fact, this comparison is yet another hard shadow calculation, which yields similar visual results to the stencil shadow method described in the introduction. Calculating a soft shadow with this shadow map is where percentage-closer filtering (**PCF**) comes into the picture.

PCF estimates the penumbra by extending the shadow test to the pixels' neighbors. By repeating the same test described earlier with the surrounding pixels, we get the sum of pixels that are closer to the light than the center pixel depth.

The following screenshot demonstrates how a single sample is taken (on the left-hand side in red) compared to PCF (on the right-hand side, that is, center in red and neighbors in blue):

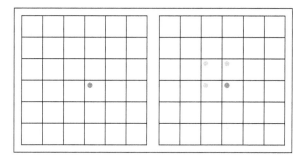

By normalizing this sum, we get the percentage of those pixels and hence the name percentage-closer filtering. If we think about the value range for this percentage, we will have the following three groups:

- Value of 0 when all the samples were in shadow (our pixel is fully in shadow).
- Value of 1 when all samples were fully lit (our pixel is fully lit).
- Any value between 0 and 1, that is, our pixel is partially in shadow that means it is in the penumbra section. The lower the value, the darker the penumbra is for that pixel.

As you can see from the percentage value range, this normalized sum provides a decent estimate for the penumbra and can be used to attenuate the corresponding light's intensity. The amount of texture pixels that get sampled during the PCF calculation are called taps. In this recipe, the hardware PCF we are using takes four taps per each call to `SampleCmpLevelZero`.

Now that we have covered the basics of PCF shadows, we can cover the shadow map generation step. Generating the matrices for the shadow map generation is very similar to how the scene's camera matrices are generated. Using those matrices, we can render the scene from the light's point of view into the shadow map and get the depth values in relation to the light source. One thing that may seem strange is the need to define a near plane. It would make more sense that the shadow range is going to start from the light source position and not from some handpicked distance value. The reason we need to define the near plane distance is so that our projection matrix could normalize the nonlinear depths before they get written into the shadow map. Choosing an appropriate value for the near plane depends on the scene being rendered. A good rule of thumb is to push the near plane as far from the light source as possible without it clipping any of the visible shadow casters. Although it is best to find a single value suitable for the entire scene, in very complex scenes you may have to tune this value for individual lights.

What are the bias values and why do we need them? Most of the problems shadow mapping suffers from are due to inconsistent mapping between the scene and shadow map pixels. The following illustration shows a scenario where a mesh surface (in red) gets rendered into the shadow map (yellow lines represent mapping from surface to pixels in the shadow map). This same surface is then rendered into GBuffer and gets lit using the scene's camera (blue lines represent mapping from the surface to the back buffer pixels):

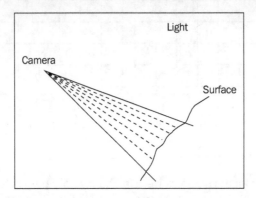

In this scenario, we can see that the surface positions of the pixels rendered into the GBuffer don't match the surface positions of the pixels rendered into the shadow map. To make this worse, the same surface area got covered by more GBuffer pixels compared to the shadow map. When the GBuffer world positions are transformed and compared with the shadow map values, the results are going to be wrong, because the positions represented by the pixels in the GBuffer and the shadow map don't match. Since the scene camera can move freely, this situation is in fact unavoidable.

Here is a side-by-side comparison of how shadows would look in a scene when those issues are not dealt with (on the left-hand side) compared to how the shadows should look (on the right-hand side):

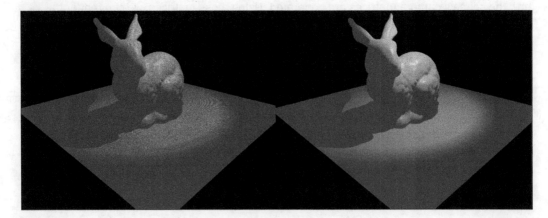

As you can see form the left image of the screenshot, there are a lot of pixels that got false shadowed. In addition, some of the shadows are a bit bulky and pixilated. This is obviously unacceptable.

A perfect and obvious solution to this problem would be to change the shadow map, so that its resolution and mapping will match exactly the one used by the scene camera. Unfortunately, this solution is not realistic because of the following reasons:

- There is a limit on how large the shadow map can be, so it will always be possible to position the scene camera in a way that will require a larger shadow map than the largest size we can allocate.

- The shadow's penumbra size depends on the amount of pixels sampled around the center pixel. When a large shadow map is used, the distance each pixel represents gets smaller, which leads to a smaller penumbra. It is always possible to take more samples around the center pixel to increase the penumbra size, but that will affect performance.

- For a light source that doesn't move, the surface to shadow map mapping is always going to be constant. Without restricting the camera movement to positions that map nicely into the shadow map, there is no way to ensure there is a one-to-one mapping between the GBuffer pixels and the shadow map pixels.

Since we can't solve this issue using the perfect solution, we have to use other solutions that reduce the problems as much as possible, so we can get the results shown on the image that is on the right-hand side. The most common solution to this false-shadowing issue is to push the depths in the shadow map towards the far plane by a small value we call a depth bias. In addition to the depth bias that pushes all the pixels by the same value, we can also use a slope scale bias that takes curvature into account. Slope scale bias pushes pixels based on the angle between the triangle and the light source direction. The more the triangle is far from facing the light, the further its pixel's depth will be pushed out.

Tuning the bias values correctly is a very delicate task. Too much bias and the shadows will detach from the casters resulting in what's known as peter panning. Not enough bias and you end up with the aliasing shown on the image that is on the left-hand side. If you find that one value is not working for all light sources, you may want to use different values based on the light properties. For example, you may find that lights with long range require higher bias values to look right. Keep in mind that there is a limit to the amount of different render states you can create, so having a separate state for each light source is probably overkill.

Choosing between the 16-bit and the 32-bit texture format, the texture format depends mostly on available memory and the shadow quality you are expecting. The 32-bit textures provide a better depth range compared to 16-bit textures, which will reduce the aliasing issues. This doesn't mean you can't get good-looking shadows with 16-bit textures, so you should experiment with both and choose the format most suitable for your scenario.

Nothing prevents you from mixing both formats and using one for some of the lights and the other for the rest of them. Just make sure you are using the appropriate bias for each one of them.

Generating a UV coordinate from a world position using the spot light's view-projection transformation was already covered in the *Projected texture – spot light* recipe of *Chapter 1, Forward Lighting*. It is recommended to go over the explanation again if you are not sure how this calculations works.

A new feature of DirectX 10 we used in this example for the first time is the comparison sampler and the use of `SampleCmpLevelZero`. Up to Version 10, DirectX texture samplers supported only point sampling and filtered sampling that would blend four taps into a single value. PCF does not work with blending for two reasons:

▸ Blending depth values sampled from the shadow map will only smooth the resulting value. As you may recall, the penumbra is a result of a large depth difference between the shadow caster and receiver. Smoothening the depths will hide this situation, which will provide false results.

▸ If we take a single sample using a filter, this sample will return a single value. When you compare that value against the receiver depth, the result is going to be Boolean. As you recall from the shadow mapping overview, this comparison will result in hard shadows, which is not what we are looking for.

DirectX 10 and above provide a sampling with comparison, which means that the hardware compares each sample first and filters the result. This is exactly how PCF works, so we no longer have to write the code that does the PCF calculation in the pixel shader. In addition, since the comparison is implemented in hardware, we get better performance.

There's more...

In this recipe, we covered the deferred light implementation, but what about forward lighting? The best way to cast shadows from a forward rendered spot light is to use the single spot light shader. As with deferred lighting, you will still need to prepare the shadow map before you do the lighting. Just change `CalcSpot` in the same way we did for the deferred lighting.

If you are using the deferred approach, it is very likely will still need forward lighting to light semitransparent geometry. It is recommended that you decide which light's cast shadow to go for and prepare their shadow map before any type of lighting is calculated. You can then use those shadow maps for both deferred and forward lighting when lighting with one of those selected light sources.

Point light PCF shadows

Casting shadows from a point light used to be difficult compared to spot lights before the days of DirectX 10. Similar to spot light shadow map, the point light shadow map has to cover the entire area affected by the light. Cube-map textures seem like a natural candidate for this task, but they could not be used properly with DirectX 9 due to two problems.

 Cube maps can be thought of as six 2D textures, one texture for each face of the cube. Cube maps are great for representing information around a center point, which in this recipe, is the point light's center.

The first problem was memory. A point light with the same range as a spot light will need six times the amount of memory to cast shadows of the same quality due to the cube map's six faces. This may have been acceptable if that memory could have been reused for both point and spot light shadows, but there was no way to do that.

Problem number two had to do with the lack of hardware PCF support. When implementing PCF in software, the sampling position for each one of the taps has to be calculated in the pixel shader. However, sampling a cube map requires a sampling direction vector, not a UV coordinate. Finding the direction for the center sample is easy, since it's just the vector from the light to the pixel getting lit. Unfortunately, finding the vector to the neighboring pixels is a different story. Finding those extra vectors will require a relatively large amount of math operations that will affect performance.

Due to these two problems, cube maps were not a popular storage medium for point light shadow mapping. Instead, multiple 2D textures were used in various ways. Among the popular solutions, split the point light into six shadow casting spot lights (same memory size as a cube map, but its reusable for regular spot lights and PCF is handled the same as for spot lights) and a solution called Praboloid Shadow Maps, which uses two 2D textures to represent the sphere (heavier on the math compared to spot light shadows, but uses third of the cube-map memory). Memory is reusable for spot lights and PCF is easy to perform.

DirectX 10 introduced new features that make cube maps attractive again for shadow mapping. Texture arrays make rendering the depths into a cube map much easier and probably faster. Shader resource views make it possible to reuse a cube map as an array of six spot light shadow maps. Finally, all the complexity of implementing PCF for cube maps is gone since PCF is implemented in hardware.

Getting ready

When rendering into a cube map, we will need a separate view matrix for each one of these faces. Those view matrices are then combined with a projection matrix with similar values to the ones used for spot light shadow, only this time the field of view is fixed to 90 degrees.

The view matrices used for each one of the cube map faces are calculated with a fixed rotation and the position of the light source transformed by the rotation. This is similar to the way we calculated the translation portion of the view matrix for spot light shadows. Here are the final view matrices for each cube face where *P* is the light position after transformed by the appropriate rotation:

$$
\text{Matrix 0}
\begin{bmatrix}
0 & 0 & -1 & 0 \\
0 & 1 & 0 & 0 \\
1 & 0 & 0 & 1 \\
P & P & P & 0
\end{bmatrix}
\quad
\text{Matrix 1}
\begin{bmatrix}
0 & 0 & 1 & 0 \\
0 & 1 & 0 & 0 \\
-1 & 0 & 0 & 0 \\
P & P & P & 1
\end{bmatrix}
\quad
\text{Matrix 2}
\begin{bmatrix}
1 & 0 & 0 & 0 \\
0 & 0 & 1 & 0 \\
0 & -1 & 0 & 0 \\
P & P & P & 1
\end{bmatrix}
$$

$$
\text{Matrix 3}
\begin{bmatrix}
1 & 0 & 0 & 0 \\
0 & 0 & -1 & 0 \\
0 & 1 & 0 & 0 \\
P & P & P & 1
\end{bmatrix}
\quad
\text{Matrix 4}
\begin{bmatrix}
1 & 0 & 0 & 0 \\
0 & 1 & 0 & 0 \\
0 & 0 & 1 & 0 \\
P & P & P & 1
\end{bmatrix}
\quad
\text{Matrix 5}
\begin{bmatrix}
-1 & 0 & 0 & 0 \\
0 & 1 & 0 & 0 \\
0 & 0 & -1 & 0 \\
P & P & P & 1
\end{bmatrix}
$$

The projection matrix is generated in the same way we did for spot light shadows, but only this time the **Field of View** (**FOV**) value is 90:

$$
\begin{bmatrix}
s & 0 & 0 & 0 \\
0 & s & 0 & 0 \\
0 & 0 & Q & 1 \\
0 & 0 & -QZ_n & 0
\end{bmatrix}
\qquad
\begin{aligned}
S &= \cot\left(\frac{fov}{2}\right) \\
Q &= \frac{Z_f}{Z_f - Z_n}
\end{aligned}
$$

Once all these matrices are ready, we will have to combine each one of the six view matrices with the projection matrix for the final per-cube face view projection transformation.

In order to generate the shadow map, we will need to render the portion of the scene affected by the point light into a depth render target. Since this time we will be using a cube map, we will need to create it with different parameters. Use the following values for the descriptor when calling `CreateTexture2D`:

Parameter	Value
MipLevels	1
ArraySize	6
Format	DXGI_FORMAT_R16_TYPELESS or DXGI_FORMAT_R32_TYPELESS
Usage	D3D10_USAGE_DEFAULT
BindFlags	D3D10_BIND_DEPTH_STENCIL | D3D10_BIND_SHADER_RESOURCE
MiscFlags	D3D10_RESOURCE_MISC_TEXTURECUBE

As with the spot light, we will need a depth stencil view for rendering into the shadow map and a resource view to sample from it. For the depth view you will need to use the same descriptor as you did with the spot light, but only the view dimension should be set to D3D10_DSV_DIMENSION_TEXTURE2DARRAY and make the array size equal to 6. For the resource view, use the same values as you did with the spot light, but only change the view dimension to D3D10_SRV_DIMENSION_TEXTURECUBE.

Sampling the cube map in the shader will be done with the same comparison filter we used for the spot light, so we get the PCF support.

You may be able to use the same rasterizer state we used for spot light shadows, but if you see that the bias value are not working for the point light, then just create a separate rasterizer state with the appropriate values.

How to do it...

We start off by clearing the depth view, setting it along with null color render targets, and setting the rasterizer state. Once everything is set, we can start render the meshes affected by the light into the shadow map. However, we will need to change the shaders a bit in order to get each mesh rendered into the appropriate cube map faces.

Instead of outputting positions in projected- space, our vertex shader has to be modified to output positions in world space. This way we can do the per-cube face view projection transformation in the geometry shader. The code for the vertex shader is as follows:

```
float4 PointShadowGenVS(float4 Pos : POSITION) : SV_Position
{
    return Pos;
}
```

Our geometry shader will be using the following constant buffer to hold the matrices in an array:

```
cbuffer cbuffercbShadowMapCubeGS : register( b0 )
{
  float4x4 CubeViewProj[6] : packoffset( c0 );
}
```

We will need a new output structure for the geometry shader, which is defined as follows:

```
struct GS_OUTPUT
{
  float4 Pos   : SV_POSITION;
  uint RTIndex : SV_RenderTargetArrayIndex;
};
```

The geometry shader will duplicate each input triangle six times and will transform each duplicated position into one of the cube face's projected space. Along with the projected position, the geometry shader will also mark each new triangle with the render target it should get rendered into. Finally, all faces will be added to the same stream. The code for the geometry shader is as follows:

```
[maxvertexcount(18)]
void PointShadowGenGS(triangle float4 InPos[3] : SV_Position, inout
TriangleStream<GS_OUTPUT> OutStream)
{
  for(int iFace = 0; iFace < 6; iFace++ )
  {
    GS_OUTPUT output;

    output.RTIndex = iFace;

    for(int v = 0; v < 3; v++ )
    {
      output.Pos = mul(InPos[v], CubeViewProj[iFace]);
      OutStream.Append(output);
    }
    OutStream.RestartStrip();
  }
}
```

This function takes the output from the vertex shader, which contains the world-space positions and outputs the six duplicated triangles in projected space into the output stream.

Now that the shadow map is ready, we can move on to the lighting calculation. Unlike spot light shadows that require a transformation to shadow space, point maps use the direction from the light source to the pixel for sampling the cube map. This means that we don't need to transform the pixel's world-space position to projected space for the UV. We will, however, need to come up with the pixel's depth value corresponding to the cube-map face this pixel belongs to. For that we will only need two of the projection parameters from the projection matrix we prepared. The updated pixel shader constant buffer is as follows:

```
cbuffer cbPointLightPixel : register( b1 )
{
```

```
    float3 PointLightPos        : packoffset( c0 );
    float PointLightRangeRcp    : packoffset( c0.w );
    float3 PointColor           : packoffset( c1 );
    float2 LightPerspectiveValues  : packoffset( c2 );
}
```

The new constant buffer member `LightPerspectiveValues` should be filled with the two last values of the third column of the perspective matrix.

In order to sample the shadow map, we will add the following texture and sampler deceleration in the pixel shader:

```
TextureCube<float> PointShadowMapTexture : register( t4
);SamplerComparisonState PCFSampler  : register( s1 );
```

Calculating the shadow attenuation value will be handled by the following function:

```
float PointShadowPCF(float3 ToPixel)
{
  float3 ToPixelAbs = abs(ToPixel);
  float Z = max(ToPixelAbs.x, max(ToPixelAbs.y, ToPixelAbs.z));
  float Depth = (LightPerspectiveValues.x * Z +
  LightPerspectiveValues.y) / Z;
  return PointShadowMapTexture.SampleCmpLevelZero(PCFSampler, ToPixel,
  Depth);
}
```

This functions the non-normalized direction from the light to the pixel and returns the shadow attenuation for the pixel being lit.

As with the spot light, you would want to support both nonshadow casting and shadow casting point lights. We will be using the same pixel shader entry point solution as we did with the spot light:

```
float4 PointLightPS( DS_OUTPUT In ) : SV_TARGET
{
  return PointLightCommonPS(In, false);
}


float4 PointLightShadowPS( DS_OUTPUT In ) : SV_TARGET
{
  return PointLightCommonPS(In, true);
}
```

And the common function is as follows:

```
float4 PointLightCommonPS( DS_OUTPUT In, boolbUseShadow )
{
    …

finalColor.xyz = CalcPoint(position, mat, bUseShadow);

    …

}
```

Now that we have the entry points and common function set, we can show the changes in `CalcPoint` to add the PCF support:

```
float3 CalcPoint(float3 position, Material material, bool bUseShadow)
{
    float3 ToLight = PointLightPos - position;
    float3 ToEye = EyePosition - position;
    float DistToLight = length(ToLight);

    // Phong diffuse
    ToLight /= DistToLight; // Normalize
    float NDotL = saturate(dot(ToLight, material.normal));
    float3 finalColor = material.diffuseColor.rgb * NDotL;

    // Blinn specular
    ToEye = normalize(ToEye);
    float3 HalfWay = normalize(ToEye + ToLight);
    float NDotH = saturate(dot(HalfWay, material.normal));
    finalColor += pow(NDotH, material.specPow) * material.specIntensity;

    float shadowAtt;
    if(bUseShadow)
    {
       // Find the shadow attenuation for the pixels world position
       shadowAtt = PointShadowPCF(position - PointLightPos);
    }
    else
    {
       // No shadow attenuation
       shadowAtt = 1.0;
    }

    // Attenuation
    float DistToLightNorm = 1.0 - saturate(DistToLight *
    PointLightRangeRcp);
```

```
    float Attn = DistToLightNorm * DistToLightNorm;
    finalColor *= PointColor.rgb * Attn * shadowAtt;

    return finalColor;
}
```

This function uses the same Boolean branch we presented in the spot shadow's implementation to skip the shadow calculations when they are not needed. Note that the shadow calculation function needs the non-normalized direction to the pixel, so the function gets called just before that vector is normalized.

How it works...

Rendering the scene into the shadow map gets easier when we treat the cube map as an array of six 2D depth targets. We render each mesh once and take advantage of the geometry shaders ability to output multiple triangles. Each input triangle is duplicated six times and gets transformed with each one of the transformations to the cube-map faces. During the shadow map generation, we used a new system value SV_RenderTargetArrayIndex. This system value lets us mark each output triangle in the geometry shader with the render target we want it to get rasterized into.

Calculating the PCF attenuation value requires a sample direction and the depth value to compare against. Similar to how we sampled the cube map in the the *Projected texture – point light* recipe of *Chapter 1, Forward Lighting*, we use the light to pixel direction (no need to normalize the vector). For the comparison depth, we need to transform the pixel's world position into the projected space of the cube-map face the direction is pointing to. In order to avoid complicated calculations to select the right transformation out of the six, we use a small trick to get the pixel's view space position for the correct cube face. All view-space transformations are axis aligned. Combined with the 90 degree FOV used by the projection matrix, we can use simple trigonometry to figure out that the Z value in the correct view space is always going to be the component of the light to pixel vector that is furthest from the origin. Note that the furthest component can be negative, so we have to compare absolute values. This is not an issue since the depth value is always going to be positive once the world-space position is fully transformed to projected space.

Once we got the Z value in the correct view space, we need to project it into the nonlinear space stored in the depth map. Since we use the same projection matrix for all cube-map faces, we can use the same projection values regardless of the cube-map face we will be sampling from. The math we use for projection is equivalent to the result we would get from multiplying the full view-space position and dividing the Z component by the W component.

Calculating the PCF attenuation value is done with the sample comparison sampler we used for the spot light shadow and gets handled in hardware for us. The resulting attenuation is then combined with the light's range attenuation, which is our final attenuation value for the pixel.

There's more...

Most meshes don't need to be rendered into all six cube-map faces. You can optimize the shadow map generation time by pretesting each mesh visibility for each one of the six transformations. We can then use an unsigned integer constant to pass a visibility bit flags to the geometry shader. In the geometry shader, the visibility flag can then be tested for each one of the six faces in the main loop. When the mesh is not visible for any of the cube faces, we can just skip outputting the triangle for the invisible cube faces.

Cascaded shadow maps

In the previous two chapters when we covered lighting calculations, the directional light was always the first recipe. The directional light has no attenuation to worry about and it spans infinitely in every direction. These two qualities which make directional lighting easy to calculate, turn into an obstacle when calculating shadows.

In order to add shadow's support to the directional light, we need a way to map the entire scene area into a fixed size shadow map. For a very small scene, it may be possible to allocate a large shadow map and have it cover the entire scene. For a decent-sized scene, however, this will result in very bad scene to shadow map pixel ratio. If you recall from the introduction, this pixel ratio is a key factor in the amount of shadow-related artefacts we are going to see in the lighting/shadowing result image. Getting good quality shadows while lighting a decent-sized scene requires a more sophisticated solution.

Since we can't cover the whole scene with a single shadow map, a different approach would be to only cover the visible portion of the scene. This solution performs better since the visible portion of the scene is bound by the projection parameters used for the scene's camera. A possible implementation would be to find an orthographic projection oriented in the directional lights direction that encapsulates the scene's visible area. We then use this orthographic projection to generate and calculate the shadows in the exact same way we did for spot light shadows.

Even though we get better pixel coverage by narrowing down the area covered by the shadow map, for a camera that defines a relatively far range, the coverage may not be good enough. This approach also introduces new problems due to the way the scene pixels get mapped to the shadow map. Each time the camera moves, the content of the shadow map changes. As explained in the introduction, we never get an exact one-to-one mapping between the scene and the shadow map. So now, a scene pixel that was mapped to a shadow map pixel with one value can get mapped to a shadow map pixel with a different value based on the camera's movement. This inconsistent mapping results in shadow shimmering. Note that the mapping can change by both, rotation and translation of the camera.

Another problem with this solution is wasted pixels due to difference between the shape of the shadow map texture and the shape of the camera's visible area. A camera using perspective projection has a visible area in the shape of a frustum. The shadow map, on the other hand, has a shape of a square. The following illustration overlays the frustum on the shadow map showing the wasted pixels in red:

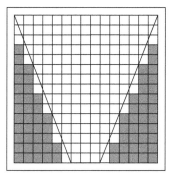

Fortunately, all these problems can be addressed with varying degrees of success. Reducing the amount of wasted pixels is partially solved by a method called cascaded shadow maps. The cascaded shadow maps reduce pixel waste by dividing the camera's visible area into multiple sections or cascades, where each cascade has its own shadow map. The following illustration features a frustum divided into three, equal-sized shadow maps:

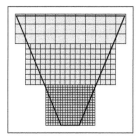

As you can see in the illustration, the pixel density is the highest at the cascade closest to the camera and it gets worse for each cascade. Although the portion of each cascade is still wasted, the overall usage is better and the pixel density is higher and close to the camera where it matters.

Selecting the camera range covered by each cascade can be difficult depending on the scene getting rendered. There are a few articles about automating the range selection based on the camera range, but I found that manual selection yields better results. Note that the cascades don't have to cover the entire view range. If you are using a camera with a large range, it would make more sense to cover a smaller portion of the range to improve the cascade's pixel coverage.

Shimmering due to camera translation and rotation is handled separately. Translation-related shimmering can be addressed by making sure the transformation into each cascade space is always done in full pixel steps. Translating in full pixel steps moves values located in the center of map without changing them. Rotation is handled by scaling the transformation to cascade space to a size large enough to encapsulate a 360 degree rotation by the camera. This extra scale prevents any changes in the shadow map when the camera rotates.

Getting ready

We will be using a 2D texture array to store the cascades. So, for three cascades, we will be using a texture array with a size of three. Unlike the point light shadow implementation, where we used a geometry shader to duplicate the triangles, this time we will be rendering each mesh only to the cascades it overlaps with. The reason for this is that the directional light shadows usually cover a larger area compared to spot and point lights. This means that we will be rendering a lot of meshes into each cascade and it would be better for performance to avoid rendering meshes to cascades they don't intersect with. In order to render each mesh separately to each cascade, you will need a separate depth view for each cascade.

It is recommended that you use 32-bit depth target for storage, as the range we are going to store into each cascade is going to be longer compared to the spot and point lights. The formats for the texture and views are as follows:

	Format
Texture	DXGI_FORMAT_R32_TYPELESS
Depth stencil view	DXGI_FORMAT_D32_UNORM
Shader resource	DXGI_FORMAT_R32_UNORM

Generating the view matrix for the cascaded shadows is similar to how we prepare the view matrix for spot light with projected texture in *Chapter 1, Forward Lighting*. Please refer to that recipe if you don't remember how to prepare the view matrix.

The directional light rays are parallel to each other, so a perspective projection is not going to work. Instead, we will be using orthographic projection. An orthographic projection with a width of *w*, height of *h*, near plane distance of *zn*, and far plane distance of *zf* is represented by the follow matrix:

$$\begin{bmatrix} 2/w & 0 & 0 & 0 \\ 0 & 2/h & 0 & 0 \\ 0 & 0 & 1/(zf\text{-}zn) & 0 \\ 0 & 0 & zn/(zn\text{-}zf) & 1 \end{bmatrix}$$

We will be using one orthographic projection to transform into a global shadow space and another orthographic projection to transform from the global shadow space into each cascade's space. Generating the global space projection requires the scene camera frustum's bounding box in the shadow view space. You can extract the frustum corners from scene camera view and projection matrix along with the near and far plane distances as follows:

- ► `RightTopNearPosition = ViewMatrix[3] + ViewMatrix[0] * NearPlane * ProjectionMatrix[0][0] + ViewMatrix[1] * NearPlane * ProjectionMatrix[1][1] + ViewMatrix[3] * NearPlane`

- ► `RightTopFarPosition = ViewMatrix[3] + ViewMatrix[0] * FarPlane * ProjectionMatrix[0][0] + ViewMatrix[1] * FarPlane * ProjectionMatrix[1][1] + ViewMatrix[3] * FarPlane`

- ► `LeftTopNearPosition = ViewMatrix[3] - ViewMatrix[0] * NearPlane * ProjectionMatrix[0][0] + ViewMatrix[1] * NearPlane * ProjectionMatrix[1][1] + ViewMatrix[3] * NearPlane`

- ► `LeftTopFarPosition = ViewMatrix[3] - ViewMatrix[0] * FarPlane * ProjectionMatrix[0][0] + ViewMatrix[1] * FarPlane * ProjectionMatrix[1][1] + ViewMatrix[3] * FarPlane`

- ► `RightBottomNearPosition = ViewMatrix[3] + ViewMatrix[0] * NearPlane * ProjectionMatrix[0][0] - ViewMatrix[1] * NearPlane * ProjectionMatrix[1][1] + ViewMatrix[3] * NearPlane`

- ► `RightBottomFarPosition = ViewMatrix[3] + ViewMatrix[0] * FarPlane * ProjectionMatrix[0][0] - ViewMatrix[1] * FarPlane * ProjectionMatrix[1][1] + ViewMatrix[3] * FarPlane`

To find the bounding box, just transform all six positions with the shadow view matrix and find the minimum and maximum values of each one of the three position components. You can now prepare the orthographic projection into the global shadow space by using bounding box width as the value *w*, height as the value *h*, depth as the value *zf*, and zero as *zn*.

Now we can find the orthographic projection of each one of the cascades. Each cascade contains a different range of the scene's camera frustum as explained in the introduction. Each cascade will have its own near and far range on the scene's camera frustum with no overlaps. Use the same calculations we used for the global shadow projection to find the six frustum points of each one of the cascades. Instead of generating the orthographic projection straight from those positions, we want to use the extra scale mentioned in the introduction to avoid rotation shimmering. To find this scale, transform the positions with the combined view and global shadow matrices and calculate the bounding sphere for each cascade group of frustum positions. Then, encapsulate the sphere with a bounding box and generate the orthographic projection from that box. Each one of those matrices is the transformation from the global shadow space to the individual cascade space.

Finally we want to avoid translation-related shimmering. We will need to keep track of the scene's camera world position individually for each cascade in order to do that. First, find the size a single pixel of each cascade in cascade space units. Each time the camera moves, you will need to transform the previous tracking position of each cascade and the new camera position all the way to each cascade space. If X or Y difference is at least one full pixel size, move the

How to do it...

We start by filling each cascade with all the meshes that intersect with its bounds as explained in the *Getting ready* section. Set the rasterizer state with the bias values and clear all the views. For each cascade, set the corresponding view and full transformation combined of the shadow view, global shadow space, and corresponding cascade. With everything set, render all the meshes intersecting with cascade bounds with the usual vertex shader and without a pixel shader. Repeat this process for each cascade.

Once all cascades have been filled, we can move on to the lighting calculations. We declare the texture array containing the shadow cascades as follows:

```
Texture2DArray<float>CascadeShadowMapTexture : register( t4 );
```

Selecting the right cascade for each pixel getting lit will require some additions to the constant buffer we use for the nonshadowed deferred and forward light. First, we need the matrix that transforms from world space to the directional shadow space defined as `ToShadowSpace`. We will also need the individual transforms from the directional shadow space to the individual cascades declared as `ToCascadeSpace`:

```
cbuffer cbDirLight : register( b1 )
{
   float3 AmbientDown : packoffset( c0 );
   float3 AmbientRange : packoffset( c1 );
   float3 DirToLight  : packoffset( c2 );
   float3 DirLightColor : packoffset( c3 );
   float4x4 ToShadowSpace : packoffset( c4 );
   float4 ToCascadeOffsetX  : packoffset( c8 );
```

```
    float4 ToCascadeOffsetY  : packoffset( c9 );
    float4 ToCascadeScale  : packoffset( c10 );
}
```

We will need to change the directional lighting function to calculate the shadow attenuation and use it with the Blinn-phong calculation. The modified function is as follows:

```
float3 CalcDirectional(float3 position, Material material)
{
  // Phong diffuse
  Float NDotL = dot(DirToLight, material.normal);
  float3 finalColor = DirLightColor.rgb * saturate(NDotL);

  // Blinn specular
  float3 ToEye = EyePosition - position;
  ToEye = normalize(ToEye);
  float3 HalfWay = normalize(ToEye + DirToLight);
  float NDotH = saturate(dot(HalfWay, material.normal));
  finalColor += DirLightColor.rgb * pow(NDotH, material.specPow) *
  material.specIntensity;

  // Shadow calculation
  float shadowAtt = CascadedShadow(position);

  return shadowAtt * finalColor * material.diffuseColor.rgb;
}
```

The function that calculates the cascade shadow attenuation is as follows:

```
float CascadedShadow(float3 position)
{
  // Transform the world position to shadow space
  float4 posShadowSpace = mul(float4(position, 1.0), ToShadowSpace);

  // Transform the shadow space position into each cascade position
  float4 posCascadeSpaceX = (ToCascadeOffsetX + posShadowSpace.xxxx) *
  ToCascadeScale;
  float4 posCascadeSpaceY = (ToCascadeOffsetY + posShadowSpace.yyyy) *
  ToCascadeScale;

  // Check which cascade we are in
  float4 inCascadeX = abs(posCascadeSpaceX) <= 1.0;
  float4 inCascadeY = abs(posCascadeSpaceY) <= 1.0;
  float4 inCascade = inCascadeX * inCascadeY;

  // Prepare a mask for the highest quality cascade the position is in
  float4 bestCascadeMask = inCascade;
```

```
        bestCascadeMask.yzw = (1.0 - bestCascadeMask.x) * bestCascadeMask.yzw;
        bestCascadeMask.zw = (1.0 - bestCascadeMask.y) * bestCascadeMask.zw;
        bestCascadeMask.w = (1.0 - bestCascadeMask.z) * bestCascadeMask.w;
        float bestCascade = dot(bestCascadeMask, float4(0.0, 1.0, 2.0, 3.0));

        // Pick the position in the selected cascade
        float3 UVD;
        UVD.x = dot(posCascadeSpaceX, bestCascadeMask);
        UVD.y = dot(posCascadeSpaceY, bestCascadeMask);
        UVD.z = posShadowSpace.z;

        // Convert to shadow map UV values
        UVD.xy = 0.5 * UVD.xy + 0.5;
        UVD.y = 1.0 - UVD.y;

        // Compute the hardware PCF value
        float shadow = CascadeShadowMapTexture.SampleCmpLevelZero(PCFSampler,
        float3(UVD.xy, bestCascade), UVD.z);

        // set the shadow to one (fully lit) for positions with no cascade
        coverage
        shadow = saturate(shadow + 1.0 - any(bestCascadeMask));

        return shadow;
    }
```

This function takes the pixel's world position and returns the attenuation as found in the closest cascade to the camera's position.

How it works...

Compared to the spot light shadows, the HLSL code for generating the shadow map is exactly the same. Using the shadow maps is also very similar except for the cascade selection. Finding the best cascade for each scene pixel is based on a 2D position intersection with a rectangle. We always want to use the cascade closest to the camera position the pixel intersects with, as that cascade will contain the most information around the pixel position for the PCF calculation.

Finding the closest cascade is straight forward. First, we transform the pixel's world position to the global shadow space. From there, we transform the position to each cascade space from the furthest cascade to the closest cascade. For each cascade, we test if the pixel position in cascade space is inside the cascade's rectangle bounds. If so, we change the cascade selection index to the index of that cascade. At the end of this process, the cascade selection index will contain the index to the closest cascade the pixel intersects with. That cascade is the one we use for the PCF calculation.

PCF with varying penumbra size

Now that we covered the basics of shadow casting with the different light sources, we can extend the fixed penumbra PCF implementation to support varying penumbra size.

We will be using a technique called **percentage-closer soft shadows** (**PCSS**). The technique will be presented through the spot light, but it can be used with point light sources. Although you can get this technique to work with directional shadows, it won't make much sense. This will be further explained in the *How it works...* section.

Getting ready

For this recipe, we will be using the exact same setup we had in the spot light shadow recipe. Note that we will need to take more samples from the shadow map in order to generate the wider penumbra areas, so it is recommended to use a larger shadow map than you would use with the fixed penumbra shadows.

If you may recall, the introduction to this chapter mentioned that the light source size affects the penumbra size. However, none of our light sources define its size. In fact, all the light sources only define the volume affected by them. To get the variable penumbra size working, we will need to use some size value that will be used in the shadow calculation. It is recommended that you choose an appropriate value for each light source by manually tuning the value.

Last thing we need for the PCSS implementation is a point sampler. In *Chapter 2, Deferred Shading*, we already created such a sampler for the GBuffer unpacking. You can reuse the same sampler for the shadow as well.

How to do it...

We will be using the exact same steps to generate the shadow map as we did for the spot light with fixed penumbra. If you plan on using lights with both fixed and varying penumbra size, you should unify the code that generates the shadow map for both types of shadows.

During the lighting, we will need a few new values in the pixel shader constant buffer. As mentioned in the *Getting Ready* section, you will need to pass in a value for the light size. In addition to that, we will also need the receptional shadow map size (one over the shadow map size). The modified constant buffer is defined as follows:

```
cbuffercbSpotLightPixel : register( b1 )
{
  float3 SpotLightPos        : packoffset( c0 );
  float  SpotLightRange      : packoffset( c0.w );
  float3 SpotLightDir        : packoffset( c1 );
  float  SpotCosOuterCone    : packoffset( c1.w );
```

```
    float3 SpotColor          : packoffset( c2 );
    float SpotCosInnerCone    : packoffset( c2.w );
    float4x4 ToShadowMap      : packoffset( c3 );
    float2 ShadowMapPixelSize : packoffset( c7 );
    float LightSize           : packoffset( c7.z );
}
```

During the PCSS calculation, we will be using the following shader constant array for the PCF filtering:

```
staticconst float2 poissonDisk[16] = {
    float2( -0.94201624, -0.39906216 ),
    float2( 0.94558609, -0.76890725 ),
    float2( -0.094184101, -0.92938870 ),
    float2( 0.34495938, 0.29387760 ),
    float2( -0.91588581, 0.45771432 ),
    float2( -0.81544232, -0.87912464 ),
    float2( -0.38277543, 0.27676845 ),
    float2( 0.97484398, 0.75648379 ),
    float2( 0.44323325, -0.97511554 ),
    float2( 0.53742981, -0.47373420 ),
    float2( -0.26496911, -0.41893023 ),
    float2( 0.79197514, 0.19090188 ),
    float2( -0.24188840, 0.99706507 ),
    float2( -0.81409955, 0.91437590 ),
    float2( 0.19984126, 0.78641367 ),
    float2( 0.14383161, -0.14100790 )
    };
```

Last thing we need to define is the point sampler that will be used for the blocker search. If the light you are adding the PCSS implementation to is a differed light, you already have a point sampler declared for the GBuffer unpacking. Use this new declaration only for forward spot light. The sampler state is declared as follows:

```
SamplerState BlockerSampler : register( s2 );
```

With the buffer, constants, and sampler set, we can look at the pixel shader changes. In order to support both PCF and PCSS, we will use the same trick we used for separating shadow casting and nonshadow casting shaders. Our common function will now take another input Boolean that gets passed to the spot calculation function:

```
float4 SpotLightCommonPS( DS_OUTPUT In, boolbUseShadow, boolbUsePCSS )
{
    ...

    finalColor.xyz = CalcSpot(position, mat, bUseShadow, bUsePCSS);

    ...
}
```

Both entry points we used before need to pass `false` for the new parameter, while a new entry point has to be added for using PCSS:

```
float4 SpotLightShadowPCSSPS( DS_OUTPUT In ) : SV_TARGET
{
  return SpotLightCommonPS(In, true, true);
}
```

Our spot light calculation will now use the new input to choose the shadow calculation it's going to use:

```
float3 CalcSpot(float3 position, Material material, bool bUseShadow,
bool bUsePCSS)
{
  ...

  float shadowAtt;
  if(bUseShadow)
  {
float4 posShadowMap = mul(position, ToShadowMap);
if(bUsePCSS)
    {
shadowAtt = SpotShadowPCSS(posShadowMap.xyz / posShadowMap.w);
    }
else
    {
shadowAtt = SpotShadowPCF(posShadowMap.xyz / posShadowMap.w);
      }
  }
  else
  {
    shadowAtt = 1.0;
  }

  ...
}
```

Finally, we can define the new function that calculates PCSS:

```
float SpotShadowPCSS(float3 tc)
{
  // Search for blockers
float avgBlockerDepth = 0;
float blockerCount = 0;
  for(inti=-2; i<= 2; i += 2)
    {
```

```
for(int j=-2; j <= 2; j += 2)
        {
            float4 d4 = g_txShadowMap.GatherRed(BlockerSampler, tc.xy,
int2(i, j));
            float4 b4 = (tc.z<= d4) ? 0.0: 1.0;

blockerCount += dot(b4, 1.0);
avgBlockerDepth += dot(d4, b4);
        }
    }

    // Check if we can early out
if(blockerCount<= 0.0)
    {
    return 1.0;
    }

    // Penumbra width calculation
    avgBlockerDepth /= blockerCount;
    float fRatio = ((tc.z - avgBlockerDepth) * LightSize) /
    avgBlockerDepth;
    fRatio *= fRatio;

    // Apply the filter
    float att = 0;
    for(inti=0; i< 16; i++)
    {
    float2 offset = fRatio * ShadowMapPixelSize * poissonDisk[i];
       att += SpotShadowMapTexture.SampleCmpLevelZero(PCFSampler, tc.xy +
       offset, tc.z);
    }

    return att / 16.0;
}
```

This function takes the pixel's shadow map UV and shadow space depth and returns the PCSS attenuation value.

How it works...

Although fixed penumbra shadow mapping can provide acceptable visuals, they will never look as good as variable size penumbra algorithms. Extending fixed penumbra algorithms such as PCF to support variable size penumbra is possible, but unfortunately this comes at the cost of performance. Due to the popularity and old age of the current generation of consoles (Xbox360 and PlayStation3), most game developers can't afford to use a performance costly shadow mapping calculation on their titles. This is why PCF shadow mapping is so popular.

While writing this book, the next generation of consoles was just around the corner and PC-only titles, such as MMOs, are getting more and more populated. Both of these changes allow the use of newer more powerful GPUs, which can support variable size penumbra algorithms.

Let's look again at the same scenario we covered before and see how variable size penumbra can be generated by a technique called PCSS. Fixed-size penumbra algorithms ignore the size of the light source, but this information is needed for generating shadows with variable size penumbra. The following illustration shows the same scene twice. As you can see, this time the light source has a size wider than a point. On the left-hand side, we show point P inside the penumbra area colored grey. On the right-hand side, we show the region which can contain shadow casters that will affect point P:

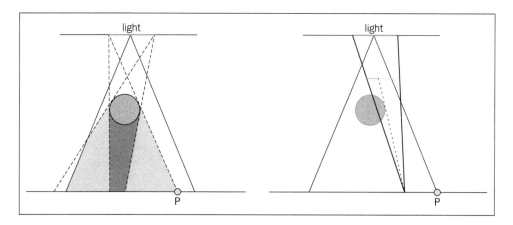

During the lighting calculations for the pixel located at point P, we will need an estimate for the penumbra size. If you look at the right-hand side of the illustration, you will see a black triangle with the light source at its base and point P at its tip. By searching for blockers inside the triangle, we can decide how much of the light area is occluded from point P. In the illustration, the red line indicates the occluded portion and the green line indicates the lit portion. By dividing the size of the red line by the combined size of the green and red lines, we can get an estimate for how wide the penumbra is in that area. This step is called the blockers search.

Now that we have a way to decide how wide the penumbra is, we need a way to use this value when doing the actual shadow calculation. Fortunately, we already know that the penumbra size for PCF shadow mapping is defined by the amount and distance between the samples. We can now use the blocker's search step result to change the distance between the PCF samples, which will control the penumbra size. The more occluded area we find in the blockers search, the wider is the gap between the PCF samples.

As you can see from the implementation, PCSS works in three stages:

- ▸ Search for blockers and early out if there are none
- ▸ Find the penumbra size
- ▸ Apply the PCF filter

While searching for blocker's use, we didn't get to use a sampling function GatherRed before. This function takes four texture samples around the given UV, which is very useful in our case, since we want to sample a large amount of neighboring pixels. Each one of the samples gets compared against the pixel's depth. Those that are closer to the light's source are potential blockers. Potential blockers are counted and their average depth is computed to estimate the penumbra size.

Note that the shadow map contains nonlinear depth values. This makes the average blocker value dependent on the camera distance from the pixel getting calculated. If you decide you want to avoid this dependency, just use the same math presented in the *GBuffer unpacking* recipe of *Chapter 2, Deferred Shading*, to convert the shadow map value to linear depth.

If no blockers are found, the pixel is going to be fully lit. When blockers are found, we proceed to the next step, which is calculating the penumbra size. This calculation approximates the penumbra size by assuming that the larger the gap between the average blocker depth and the pixel depth, and the wider the light, the wider the penumbra is going to be. Note that if we made it to this point, the blocker's average depth is always going to be positive, so the ratio is also going to be positive.

With the penumbra size calculated, we move on to calculating the PCF shadow value. Unlike the previous recipes using PCF, where we could get around with only a single sample from the shadow map, this time we are taking 16 samples. We then use the penumbra size to push the sampling positions away from the pixel UV position in the shadow map space. For the smallest penumbra size, which is zero, we end up taking all samples in the pixel location. As the penumbra size increases, the samples get further away from each other and the penumbra area expends.

When taking so many samples in a square pattern for PCF calculation, it's not unusual to end up with a wide penumbra that shows the sampling pattern. One way to counter this issue is to change the PCF sampling to a more jigged or random pattern. In this recipe we used the Poisson disk with its semirandom values to offset the sampling positions.

There's more...

The results we got from the Poisson disk are not bad, but they can be further improved on the cost of performance. A common way is to generate a per-pixel random rotation and apply it to the disk positions. This breaks the Poisson disk pattern even more by converting the smooth penumbra into a noisier pattern.

An easy way to implement this rotation is to generate a dual-channel texture containing the sine and cosine of a per-texture pixel random angle (you will need a floating point texture for that). In the pixel shader, just sample the texture and apply the 2D rotation on the Poisson disk.

Visualizing shadow maps

At one point or another, you may find yourself in a situation where one or more lights show some sort of shadow corruption. In this situation it is very helpful to see what is stored in the shadow map in order to determine if the source of the corruption is in the shadow map generation step or in the lighting step.

Shadow map's value range is higher than the 8 bit per pixel that your monitor can display. When just outputting the depth value stored in a shadow map, you are very likely to end up with a gray image that has very low contrast values. In most cases, it will be hard to figure out what is in the shadow map with this form of visualization.

One way to improve the contrast of the shadow map content is to highlight the silhouettes of the shadow map content, which will result in a nice visual outline around the objects. In this recipe we will implement this kind of visualization mode using a very simple technique.

Getting ready

All we will need for this recipe is the shadow map and a new shader used to visualize the shadow map content. Similar to the way we handled directional light, we will expend a vertex into a full screen quad, so there is no need for a vertex buffer or deceleration.

We will be using point sampler to sample the shadow map depth values. In addition, we will be using a scale and offset values to control the silhouettes, which can be tuned while visualizing the shadow map to highlight deferent object sizes.

How to do it...

We will start off by setting the topology to D3D_PRIMITIVE_TOPOLOGY_TRIANGLESTRIP and calling a draw with four vertices, which we will expend into a full screen quad. The following constants will be used for the quad's positions and UV coordinates:

```
staticconst float2 arrPos[4] = {
  float2(1.0, 1.0),
  float2(1.0, -1.0),
  float2(-1.0, 1.0),
  float2(-1.0, -1.0),
};
```

```
staticconst float2 arrUV[4] = {
  float2(1.0, 0.0),
  float2(1.0, 1.0),
  float2(0.0, 0.0),
  float2(0.0, 1.0),
};
```

Our vertex shader will use the following output structure:

```
struct VS_OUTPUT
{
  float4 Position  : SV_Position; // vertex position
  float2 UV        : TEXCOORD0;   // vertex texture coords
};
```

We will use the following vertex shader code to output the full screen quad's values:

```
VS_OUTPUT GBufferVisVS(uintVertexID : SV_VertexID )
{
    VS_OUTPUT Output;

    Output.Position = float4(arrPos[VertexID].xy, 0.0, 1.0);
    Output.UV = arrUV[VertexID].xy;

    return Output;
}
```

The pixel shader used to output the depth values with the silhouette will sample from the following texture with a point sampler:

```
Texture2D<float> ShadowMapTexture : register( to );
SamplerState PointSampler : register( s0 );
```

We will be using the following constant buffer to store the depth scale and offset values, where the scale is stored in the first component and the offset is stored in the second component:

```
Cbuffercb ShadowMapVisPS : register( b0 )
{
  float2 DepthScaleOffset : packoffset( c0 );
}
```

Finally, the code used for the pixel shader is as follows:

```
float4 GBufferVisPS ( VS_OUTPUT In ) : SV_TARGET
{
  float4 depth = ShadowMapTexture.Gather(PointSampler, In.UV.xy);
  float DDepth = dot(abs(depth.xxx - depth.yzw), float3(1.0, 1.0,
  1.0));
  float finalDepth = depth.x + DepthScaleOffset.x * saturate(DDepth -
  DepthScaleOffset.y);

  return float4(finalDepth, finalDepth, finalDepth, 1.0);
}
```

This function samples the depths and adds the silhouettes on top of them.

How it works...

Calculating the silhouette value is very simple. First, we gather the four depth values around the UV coordinate. By summing the absolute difference between the first pixel and its neighbors, we get an estimate to the amount of discontinuity in the depth values around the first pixel. Since edges always appear around a discontinuity, the larger this absolute difference is, the more likely it is that we found an edge.

Using the scale and offset, we can tune the amount of discontinuity we want to consider as an edge and how bright we want that edge to show on the screen. By tuning these two values, you can choose whether you want to highlight fine details with low discontinuity (small offset) or only focus on the larger details (larger offset).

Why only use this visualization on shadow maps? You may find it useful to use the same type of silhouette detection when visualizing any other texture that contains depth values. For example, the GBufferdepth map.

4
Postprocessing

In this chapter we will cover various postprocessing techniques:

- ▸ HDR rendering
- ▸ Adaptation
- ▸ Bloom
- ▸ Distance depth of field
- ▸ Bokeh

Introduction

In the previous chapters we covered shadow and lighting techniques. By using those techniques, you should now be able to render your 3D scene and achieve decent output image. Now it's time to take this decent output and make it look even better using a set of effects known as postprocessing.

The term postprocessing is taken from the film industry and basically refers to any changes made to the film after it was recorded. When used in relation to 3D rendering, postprocessing generally refers to a set of 2D effects applied to the lit 3D scene which enhance the final image displayed on the screen.

Aside from image enhancement, postprocessing can also dramatically change the overall look of the final image. It's not uncommon for games to use postprocessing to alter the *mood* of the 3D scene by desaturating colors for a gloomier look or by causing parts of the final image to stand out.

Another set of effects common to postprocessing are fake lens effects. Real life cameras suffer from all sorts of imperfections due to the way their lens work. By faking those imperfections, your game can achieve a more cinematic look. In recent years, this cinematic look has become so popular that nowadays every game is expected to show those cinematic imperfections.

As you will soon realize, some of the effects covered in this chapter are very important for image quality while others may only be suitable when aiming for a certain look. It's up to you to choose the set of effects that best suits your game or application needs.

HDR rendering

In the previous chapters, we have used an 8 bit per-channel texture for accumulating the lighting calculation results. The main drawback of using this format is that all output values get clamped to the range 0 to 1. On top of that, the precision of values an 8 bit per-channel texture can store is very limited. From this point on we will refer to 8 bit per-channel textures as low dynamic range textures, or LDR for short.

When lighting a bright scene, the accumulated light value may exceed the maximum LDR value of 1. Those bright areas then get clamped to the value of 1 and all the bright details get lost. We call this saturation, because the bright areas saturate the value stored in the LDR texture. To make things worse, dark scenes will result in low values. Due to the low precision offered by LDR textures, this will also result in data loss, but this time due to values getting rounded. These type of data loss may lead to the scene looking black or to banding.

Although it may be tempting to try and tune the scenes lighting in order to avoid the above issues, in large complex scenes it is impossible to find a lighting configuration that will work for every possible camera angle and position. Instead, it would make more sense to accumulate the scene's lighting into a texture format that provides better precision and doesn't get clamped to 1. Those higher precision values can be then mapped to the screen's LDR range using a technique called tone mapping with as little detail lost as possible. This recipe offers a solution to the LDR issues called High Dynamic Range rendering or HDR for short.

Getting ready

In order to support HDR rendering, we are going to replace the previously used light accumulation 8 bit per-channel texture format with an HDR texture format. DirectX11 offers a selection of texture formats that store between 10 to 32 bits per channel for us to choose from. 16 bit per channel provides a good compromise between precision and memory consumption, so you should consider using the format **DXGI_FORMAT_R16G16B16A16_FLOAT**. As always, you should experiment and compare the visual results with other available formats.

We will be using a mixture of compute and pixel shaders for the HDR to LDR tone mapping. The tone mapping technique we will be using requires two buffers, two constant buffers and two compute shaders which will be used for the scenes luminance calculation.

The first buffer will be used for intermediate luminance values storage. This buffer's memory size is calculated based on the resolution of the back buffer. Use the following parameters to create the buffer:

Buffer descriptor parameter	Value
BindFlags	D3D11_BIND_UNORDERED_ACCESS | D3D11_BIND_SHADER_RESOURCE
StructureByteStride	4 (size of float)
ByteWidth	4 * \<Total back buffer pixels> / (16 * 1024)
MiscFlags	D3D11_RESOURCE_MISC_BUFFER_STRUCTURED

All other values in the descriptor should be zero

Once this buffer is allocated, we will need a corresponding UAV (unordered access view) and SRV (shader resource view) so we can read and write to this buffer. Use the following parameters in the UAV descriptor:

UAV descriptor parameter	Value
Format	DXGI_FORMAT_UNKNOWN
ViewDimension	D3D11_UAV_DIMENSION_BUFFER
Buffer.NumElements	\<Total back buffer pixels> / (16 * 1024)

All other values in the descriptor should be zero.

Use the following parameters in the SRV descriptor:

SRV descriptor parameter	Value
Format	DXGI_FORMAT_UNKNOWN
ViewDimension	D3D11_SRV_DIMENSION_BUFFER

All other values in the descriptor should be zero.

The second buffer will store the average luminance value, which is stored as a single float. This buffers descriptor only differs from the first buffer in its size. Use the following parameters to create the buffer:

Buffer descriptor parameter	Value
BindFlags	D3D11_BIND_UNORDERED_ACCESS | D3D11_BIND_SHADER_RESOURCE
StructureByteStride	4 (size of float)
ByteWidth	4
MiscFlags	D3D11_RESOURCE_MISC_BUFFER_STRUCTURED

All other values in the descriptor should be zero.

Similar to the first buffer, we need to create a corresponding UAV and SRV so we can read and write to this buffer. Use the following parameters for the UAV:

UAV descriptor parameter	Value
Format	DXGI_FORMAT_UNKNOWN
ViewDimension	D3D11_UAV_DIMENSION_BUFFER
Buffer.NumElements	1

All other values in the descriptor should be zero.

Use the exact same parameters for the SRV as you did for the previous buffer SRV.

Finally, we will need two constant buffers, both with a size of 16 bytes. Use the following descriptor parameters for these constant buffers:

Constant buffer descriptor parameter	Value
Usage	D3D11_USAGE_DYNAMIC
BindFlags	D3D11_BIND_CONSTANT_BUFFER
CPUAccessFlags	D3D11_CPU_ACCESS_WRITE
ByteWidth	4

All other values in the descriptor should be zero.

How to do it...

All the steps taken to fill the GBuffer and light the scene should be executed as before, only now we are going to use the new texture format for the light accumulation. Once all lighting calculations have been executed, we will need to do the tone mapping to convert the HDR light results into LDR results that can be displayed on the screen.

First step is to calculate the HDR buffer luminance value. For this step we need to prepare the constant buffer and execute the two compute shaders. Our constant buffer is declared as follows:

```
cbuffer DownScaleConstants : register( b0 )
{
    // Resolution of the down scaled target: x - width, y - height
        uint2 Res   : packoffset( c0 );
    // Total pixel in the downscaled image
        uint Domain   : packoffset( c0.z );

    // Number of groups dispached on the first pass
        uint GroupSize   : packoffset( c0.w );
}
```

Set the constant buffer values as follows:

- `Res`: Back buffer width and height divided by 4
- `Domain`: Back buffer width times height divided by 16
- `GroupSize`: Back buffer width times height divided by 16 times 1024

Once the constant buffer is ready, set the HDR SRV and the intermediate luminance UAV to the compute shader. Both views are declared in the compute shader as follows:

```
Texture2D HDRTex : register( t0 );
RWStructuredBuffer<float> AverageLum : register( u0 );
During the compute shader exection, we will be using the
following thread group shared memory for storing intermediate
results:groupshared float SharedPositions[1024];
```

We will be using the following fixed value for luminance calculations:

```
static const float4 LUM_FACTOR = float4(0.299, 0.587, 0.114, 0);
```

The compute shader code is split into three helper functions and the shader entry point function. First helper function does an initial 4 x 4 downscale on each thread using the following code:

```
float DownScale4x4(uint2 CurPixel, uint groupThreadId)
{
    float avgLum = 0.0;

    // Skip out of bound pixels
    if(CurPixel.y < Res.y)
    {
        // Sum a group of 4x4 pixels
        int3 nFullResPos = int3(CurPixel * 4, 0);
```

```
        float4 downScaled = float4(0.0, 0.0, 0.0, 0.0);
        [unroll]
        for(int i = 0; i < 4; i++)
        {
            [unroll]
            for(int j = 0; j < 4; j++)
            {
                downScaled += HDRTex.Load( nFullResPos, int2(j, i) );
            }
        }
        downScaled /= 16.0;

        // Calculate the lumenace value for this pixel
        avgLum = dot(downScaled, LUM_FACTOR);

        // Write the result to the shared memory
        SharedPositions[groupThreadId] = avgLum;
    }

    // Synchronize before next step
    GroupMemoryBarrierWithGroupSync();

    return avgLum;
}
```

Our second helper function continues the downscale all the way down to four values using the following code:

```
float DownScale1024to4(uint dispatchThreadId, uint groupThreadId,
float avgLum)
{
    // Expend the downscale code from a loop
    [unroll]
    for(uint groupSize = 4, step1 = 1, step2 = 2, step3 = 3;
    groupSize < 1024;
    groupSize *= 4, step1 *= 4, step2 *= 4, step3 *= 4)
    {
        // Skip out of bound pixels
        if(groupThreadId % groupSize == 0)
        {
            // Calculate the luminance sum for this step
            float stepAvgLum = avgLum;
            stepAvgLum += dispatchThreadId+step1 < Domain ?
            SharedPositions[groupThreadId+step1] :
            avgLum;
```

```
stepAvgLum += dispatchThreadId+step2 < Domain ?
SharedPositions[groupThreadId+step2] :
avgLum;
stepAvgLum += dispatchThreadId+step3 < Domain ?
SharedPositions[groupThreadId+step3] :
avgLum;

// Store the results
avgLum = stepAvgLum;
SharedPositions[groupThreadId] = stepAvgLum;
}

// Synchronize before next step
GroupMemoryBarrierWithGroupSync();
}

return avgLum;
}
```

Finally, the four values are down scaled into a single value averaged and stored using the following code:

```
void DownScale4to1(uint dispatchThreadId, uint groupThreadId, uint
groupId, float avgLum)
{
    if(groupThreadId == 0)
    {
        // Calculate the average lumenance for this thread group
        float fFinalAvgLum = avgLum;
        fFinalAvgLum += dispatchThreadId+256 < Domain ?
        SharedPositions[groupThreadId+256] : avgLum;
        fFinalAvgLum += dispatchThreadId+512 < Domain ?
        SharedPositions[groupThreadId+512] : avgLum;
        fFinalAvgLum += dispatchThreadId+768 < Domain ?
            SharedPositions[groupThreadId+768] : avgLum;
        fFinalAvgLum /= 1024.0;

        // Write the final value into the 1D UAV which
        //  will be used on the next step
        AverageLum[groupId] = fFinalAvgLum;
    }
}
```

These are going to be used by the following shader entry point:

```
[numthreads(1024, 1, 1)]
void DownScaleFirstPass(uint3 groupId : SV_GroupID,
uint3 dispatchThreadId : SV_DispatchThreadID,
uint3 groupThreadId : SV_GroupThreadID)
{
    uint2 CurPixel = uint2(dispatchThreadId.x % Res.x,
    dispatchThreadId.x / Res.x);

    // Reduce a group of 16 pixels to a single pixel and store in the
    shared memory
    float avgLum = DownScale4x4(CurPixel, groupThreadId.x);

    // Down scale from 1024 to 4
    avgLum = DownScale1024to4(dispatchThreadId.x, groupThreadId.x,
    avgLum);

    // Downscale from 4 to 1
    DownScale4to1(dispatchThreadId.x, groupThreadId.x, groupId.x,
    avgLum);
}
```

We will dispatch this compute shader with a group count on the X dimension of: `<Total back buffer pixels> / (16 * 1024)`

Once the first compute shader has been executed, we will be executing the second compute shader with the same constant buffer, but with the intermediate luminance SRV and average luminance UAV set to the compute shader. The second compute shader code is as follows:

```
#define MAX_GROUPS 64

// Group shared memory to store the intermediate results
groupshared float SharedAvgFinal[MAX_GROUPS];

[numthreads(MAX_GROUPS, 1, 1)]
void DownScaleSecondPass(uint3 groupId : SV_GroupID, uint3
groupThreadId : SV_GroupThreadID,
    uint3 dispatchThreadId : SV_DispatchThreadID)
{
    // Fill the shared memory with the 1D values
    float avgLum = 0.0;
    if(dispatchThreadId.x < GroupSize)
    {
        avgLum = AverageValues1D[dispatchThreadId.x];
    }
```

```
SharedAvgFinal[dispatchThreadId.x] = avgLum;

GroupMemoryBarrierWithGroupSync(); // Sync before next step

// Downscale from 64 to 16
if(dispatchThreadId.x % 4 == 0)
{
   // Calculate the luminance sum for this step
   float stepAvgLum = avgLum;
   stepAvgLum += dispatchThreadId.x+1 < GroupSize ?
   SharedAvgFinal[dispatchThreadId.x+1] : avgLum;
   stepAvgLum += dispatchThreadId.x+2 < GroupSize ?
   SharedAvgFinal[dispatchThreadId.x+2] : avgLum;
   stepAvgLum += dispatchThreadId.x+3 < GroupSize ?
   SharedAvgFinal[dispatchThreadId.x+3] : avgLum;

   // Store the results
   avgLum = stepAvgLum;
   SharedAvgFinal[dispatchThreadId.x] = stepAvgLum;
}

GroupMemoryBarrierWithGroupSync(); // Sync before next step

// Downscale from 16 to 4
if(dispatchThreadId.x % 16 == 0)
{
   // Calculate the luminance sum for this step
   float stepAvgLum = avgLum;
   stepAvgLum += dispatchThreadId.x+4 < GroupSize ?
   SharedAvgFinal[dispatchThreadId.x+4] : avgLum;
   stepAvgLum += dispatchThreadId.x+8 < GroupSize ?
   SharedAvgFinal[dispatchThreadId.x+8] : avgLum;
   stepAvgLum += dispatchThreadId.x+12 < GroupSize ?
   SharedAvgFinal[dispatchThreadId.x+12] : avgLum;

   // Store the results
   avgLum = stepAvgLum;
   SharedAvgFinal[dispatchThreadId.x] = stepAvgLum;
}

GroupMemoryBarrierWithGroupSync(); // Sync before next step

// Downscale from 4 to 1
if(dispatchThreadId.x == 0)
{
```

```
            // Calculate the average luminace
            float fFinalLumValue = avgLum;
            fFinalLumValue += dispatchThreadId.x+16 < GroupSize ?
            SharedAvgFinal[dispatchThreadId.x+16] : avgLum;
            fFinalLumValue += dispatchThreadId.x+32 < GroupSize ?
            SharedAvgFinal[dispatchThreadId.x+32] : avgLum;
            fFinalLumValue += dispatchThreadId.x+48 < GroupSize ?
            SharedAvgFinal[dispatchThreadId.x+48] : avgLum;
            fFinalLumValue /= 64.0;

            AverageLum[0] = fFinalLumValue;
        }
    }
```

Now that we calculated the average luminance value, we can use it to complete the tone mapping. This time we will be rendering a full screen quad that will output the LDR values. Set the same LDR render target view we used before for light accumulation (no need to clear the color). In addition, set the HDR and average luminance SRVs so we can use those as inputs to the pixel shader.

We will be using the same technique to render the full screen quad we used for the directional light. Just bind a NULL input layout index and vertex buffers. Call `Draw` with four vertices to render the full screen quad with the following vertex shader:

```
static const float2 arrBasePos[4] = {
    float2(-1.0, 1.0),
    float2(1.0, 1.0),
    float2(-1.0, -1.0),
    float2(1.0, -1.0),
};

static const float2 arrUV[4] = {
    float2(0.0, 0.0),
    float2(1.0, 0.0),
    float2(0.0, 1.0),
    float2(1.0, 1.0),
};

struct VS_OUTPUT
{
    float4 Position   : SV_Position; // vertex position
    float2 UV         : TEXCOORD0;
};

VS_OUTPUT FullScreenQuadVS( uint VertexID : SV_VertexID )
{
```

```
    VS_OUTPUT Output;

    Output.Position = float4( arrBasePos[VertexID].xy, 0.0, 1.0);
    Output.UV = arrUV[VertexID].xy;

    return Output;
}
```

The pixel shader we will use for tone mapping declares the following resources:

```
Texture2D<float4> HDRTexture    : register( t0 );
StructuredBuffer<float> AvgLum  : register( t1 );
```

We will be using a point sampler to sample the HDR texture which is defined as follows:

```
SamplerState PointSampler : register( s0 );
```

In addition, our pixel shader will be using the following constant buffer:

```
cbuffer FinalPassConstants : register( b0 )
{
    float MiddleGrey  : packoffset( c0 );
    float LumWhiteSqr : packoffset( c0.y );
}
```

Please refer to the *How it works...* section for details about the values you should be setting these constants to.

Tone mapping the HDR values is handled by the following function:

```
float3 ToneMapping(float3 HDRColor)
{
    // Find the luminance scale for the current pixel
    float LScale = dot(HDRColor, LUM_FACTOR);
    LScale *= MiddleGrey / AvgLum[0];
    LScale = (LScale + LScale * LScale / LumWhiteSqr) / (1.0 + LScale);

    // Apply the luminance scale to the pixels color
    return HDRColor * LScale;
}
```

Finally, the pixel shader entry point is as follows:

```
float4 FinalPassPS( VS_OUTPUT In ) : SV_TARGET
{
    // Get the color sample
    float3 color = HDRTex.Sample( PointSampler, In.UV.xy ).xyz;
```

```
    // Tone mapping
    color = ToneMapping(color);

    // Output the LDR value
    return float4(color, 1.0);
}
```

The LDR output is now ready to be displayed on screen.

How it works...

Before we can explain how the tone mapping implementation works, it's important to understand the goal we are trying to achieve when converting the accumulated HDR light values into LDR values that can be displayed on the screen. Though it's possible to normalize the HDR values and use the result as the LDR values, this solution will not improve the final image by much unless the HDR values are evenly distributed. In reality, it's possible to end up in a situation where a small portion of the HDR pixels are relatively dark while most of the pixels are relatively bright. Normalizing the pixel values in this scenario will result in an image with the dark pixels almost black while the bright pixels are very close to white. This is obviously not the result we are looking for.

Instead of normalizing the values, our tone mapping approach scales the HDR values based on two parameters: Middle Gray and White. The following formula takes an HDR pixel color and converts it to its corresponding LDR value:

$$L_S = \frac{MG}{L_A}$$

$$C_{LS} = \frac{L_S + \dfrac{L_S^2}{W^2}}{1 + L_S}$$

Where:

- ► MG is the middle grey value
- ► LA is the average luminance
- ► LS is the scaled luminance
- ► W is white
- ► CLS is the color luminance scale that scales the HDR value to LDR

What this formula does is scale the pixel's luminance value to our desired middle luminance value, then scale it to the range 0 to 1 where the given scaled white value is mapped to 1. Though simple, this formula lets us control which portion of the HDR range is going to be mapped to the LDR range, using the scaled white value, in addition to letting us choose the range we are most interested in, using the middle gray value.

Now that we understand how our tone mapping algorithm works, let's look at the implementation. As you can see, our implementation is a combination of two steps: the luminance downscale which ends up with the average luminance value and the final pass where we perform the tone mapping.

Before compute shaders were available, downscaling the luminance had to be implemented using pixel shaders. A common solution was to downscale the luminance values to a single pixel in multiple intermediate steps, each step would down scale its input to a quarter of its size. Compute shaders provide a key feature that is not available in other shaders which helps both simplify and optimize this operation, shared memory.

Compute shaders execute groups of threads. Shared memory, as the name implies, is a chunk of memory which provides fast accessibility to all the threads that belong to the same group. With the possibility to share data between the threads, a group of threads can now down scale a large group of pixels into a single pixel. Unfortunately, DirectX11 has various restrictions that prevent us from down scaling the entire HDR texture to the single average luminance value:

- Each thread group is limited to 1024 or less threads
- Shared memory is limited to 16Kb
- Each thread can only write to 256 bytes of the shared memory

Our implementation downscales 16 pixels per thread using 1024 threads per group, which is the limit, we still can't downscale the full sized HDR texture to a single value in one step. These days, a game running at 720p, which is 1280x720, is very standard. For this scenario, we will need 57 thread groups to downscale the HDR luminance values to the intermediate buffer (each thread group downscales 16 * 1024 pixels. For 720p we need 56.25 threads so we round up to 57).

If you look at the first computed shader, you will notice that we first downscale the HDR values to 1/16 of the original size. This step is important as we will eventually store this downscaled HDR image for other postprocessing operations. Once this initial downscale operation is completed, we use the luminance factors to convert the pixel color to its corresponding luminance value. As you can see from the values, each color channel gets a different scale based on our eyes sensitivity to each color.

With the luminance value calculated for each thread, the downscale operation continues with multiple of 4 pixel summation steps taking down the 1024 luminance values to a single value. In each one of these steps, a three quarters of the threads stop contributing to the downscale, while the rest of the threads sum their value with the values of 3 of their closest excluded threads. At the end of each step, the threads get synchronized before the next step takes place. It is important to note that the last thread group may not have 1024 pixels to process. In fact, in the 720p scenario, the last group will only have 256 pixels to process. Our implementation goes around this issue by replacing those missing values with an existing value. Though not mathematically accurate, this approximation should be good enough. Once our thread group is down to the sum of all pixels processed by the group, we average the sum and store it for the next compute shader.

The second compute shader will downscale the intermediate results from the first shader to get the final average luminance value. We dispatch a single group for this shader, so we can now use the same summation and average technique from the first shader to come up the average luminance. We will be dividing by the average luminance, so we have to make sure its value is higher than zero.

Once we found the average luminance, we proceed to execute a full screen quad which will handle the tone mapping. Our pixel shader executes the same formulas presented earlier to scale each pixel's HDR value to its corresponding LDR value. Both Middle Gray and White values should be art driven and may need to be changed for different scenes, weather settings, or even day/night time. Before setting the Middle Gray and White values to the constant buffer, make sure to multiply the White value with the Middle Gray value and square the result as shown in the tone mapping formula.

There's more...

The tone mapping technique presented in this recipe offers decent results, but there are other ways to perform tone mapping. Tone mapping plays a crucial role in the overall look of your rendered scene, so it is very likely that you will eventually want to use a more customized algorithm which will provide the right tuning for the look you want to achieve.

Adaptation

Now that we added HDR rendering support, you may have noticed that our average luminance value is very unstable when the content of the HDR image changes due to camera movement or changes in the scene. This instability results in noticeable changes to the tone mapping range which changes the brightness of the LDR image.

When we move from a bright to a dark area, our eyes need some time to gradually adapt to these changes. This recipe adds a similar adaptation to our average luminance calculation to smooth those kinds of transitions.

Getting ready

Adding adaptation will require an additional buffer to store the previous frame's luminance. This buffer should have the exact same descriptor values as the average luminance buffer we added in the preceding recipe. Similar to the average luminance buffer, the new buffer will also require a UAV and SRV. On each frame we render, we will store the average luminance into one of the buffers and read the previous frames average luminance from the other. Once the compute shader execution has been completed, we need to switch the two buffers, so the current frame's average luminance will be used as the previous average luminance on the next frame.

In addition to the new buffer, we will also need to increase the size of the constant buffer used for the luminance downscale from 16 byte to 32 byte. See the *How it works...* section for details on the values you should set to these extra constant buffer parameters.

How to do it...

All our changes will focus on the second compute shader that calculates the final average luminance value. The deceleration for the SRV containing the previous frame's average luminance is as follows:

```
StructuredBuffer<float> PrevAvgLum : register( t1 );
```

The extended constant buffer used in this compute shader now has an additional float value:

```
cbuffer DownScaleConstants : register( b0 )
{
    uint2 Res : packoffset( c0 );   // Resolution of the qurter size target: x - width, y - height
    uint Domain : packoffset( c0.z ); // Total pixel in the downscaled image
    uint GroupSize : packoffset( c0.w ); // Number of groups dispached on the first pass
    float Adaptation : packoffset( c1 );   // Adaptation factor
}
```

Finally, replace the following line of code:

```
AverageLum[0] = fFinalLumValue;
```

With the following lines:

```
// Calculate the adaptive luminance
float fAdaptedAverageLum = lerp(PrevAvgLum[0], fFinalLumValue, Adaptation);

// Store the final value
AverageLum[0] = max(fAdaptedAverageLum, 0.0001);
```

How it works...

Adding adaptation support is relatively easy as you can see from the implementation. Instead of storing the current frame's average luminance as we did in the preceding recipe, we offset the previous frames average luminance towards the calculated value by the adaptation amount. Our eyes' adaptation time is relatively long, so it wouldn't make sense to choose a very low adaptation value. It would make sense to use a value in the range of 1 to 5 seconds. As this calculation is frame rate dependent, you can use the following formula to come up with the appropriate adaptation value:

$$\text{Max}\left[\frac{\Delta t}{\% \text{ Per Second}}, 1\right]$$

Where Δt is the time elapsed from the last rendered frame and **% Per Second** is the percentage per-second of the average luminance we want to blend with the previous average luminance. Note that this function becomes unstable as the percentage reaches zero, so you probably want to set the adaptation to 1 for relatively low percentage values.

There's more...

It is important to handle two common edge cases where we don't have a luminance history to work with. Those edge cases are the first frame and camera teleportation. On the first frame, we didn't calculate any previous frames average luminance yet, so we will be using whatever value the buffers where initialized to. This means that for a couple of frames we will be interpolating between the wrong value and the current luminance. Camera teleportation causes a similar problem. The average luminance value stored in the previous frame was calculated in a different location and may be very far from the value where the camera was teleported to. The solution for both problems is the same – use an adaptation value of 1 in order to avoid using the value stored in the previous frame.

Bloom

Bloom is another postprocessing effect commonly used with HDR which reproduces a video camera artifact. You can think of bloom as light leakage from very bright pixels to dimmer neighbouring pixels. The following images show the same scene without bloom on the left and with bloom on the right:

Bloom is tightly connected to HDR rendering as it requires the average luminance to determine which pixels are bright enough to leak into their neighbors. In addition, blooming in LDR values is very likely to saturate values which will result in an output image which looks burned.

Getting ready

First we need to add three new textures with a size of 1/16 of the HDR texture along with corresponding SRVs and UAVs. Make sure you use the same format as the HDR texture for both of these textures. The new textures will contain the following data:

- Down scaled HDR texture
- Temporary storage for the non-filtered bloom values
- Final bloom values

How to do it...

Our first task is to downscale the HDR texture. Fortunately, we already do that in the first luminance downscale pass. Add the following UAV deceleration to the compute shader:

```
RWTexture2D<float4> HDRDownScale : register( u1 );
```

You will need to set the downscale UAV along with the intermediate luminance UAV.

As our compute shader already downscales the HDR texture to 1/16 of its original size before it converts the colors to luminance values, all we need to do is just store the downscaled values into the new UAV. All you need to do is search for the following code in the compute shader:

```
downScaled /= 16.0; // Average
avgLum = dot(downScaled, LUM_FACTOR);
```

And add the code to store the downscaled value between those two lines to get the following code:

```
downScaled /= 16.0; // Average
HDRDownScale[CurPixel.xy] = downScaled; // Store the qurter resolution
image
avgLum = dot(downScaled, LUM_FACTOR); // Calculate the lumenace value
for this pixel
```

Now that we stored the downscaled values, we will be executing a compute shader that extracts the pixels bright enough to be part of the bloom. This compute shader should be dispatched after the average luminance is fully calculated. We will be using the same constant buffer we used for the luminance calculations. Change the constant buffer by adding the bloom threshold value at the end as follows:

```
cbuffer DownScaleConstants : register( b0 )
{
    uint2 Res  : packoffset( c0 );   // Resolution of the qurter size
    target: x - width, y - height
    uint Domain  : packoffset( c0.z ); // Total pixel in the downscaled
    image
    uint GroupSize  : packoffset( c0.w ); // Number of groups dispached
    on the first pass
    float Adaptation : packoffset( c1 );   // Adaptation factor
    float fBloomThreshold : packoffset( c1.y ); // Bloom threshold
    percentage
}
```

The compute shader we will be using to find the bloom values declares the following resources:

```
Texture2D<float4> HDRDownScaleTex : register( t0 );
StructuredBuffer<float> AvgLum : register( t1 );
RWTexture2D<float4> Bloom : register( u0 );
```

The code for the compute shader is as follows:

```
[numthreads(1024, 1, 1)]
void BrightPass(uint3 dispatchThreadId : SV_DispatchThreadID)
{
    uint2 CurPixel = uint2(dispatchThreadId.x % Res.x,
    dispatchThreadId.x / Res.x);

    // Skip out of bound pixels
    if(CurPixel.y < Res.y)
    {
        float4 color = HDRDownScaleTex.Load(int3(CurPixel, 0));
        float Lum = dot(color, LUM_FACTOR);
        float avgLum = AvgLum[0];

        // Find the color scale
        float colorScale = saturate(Lum - avgLum * fBloomThreshold);

        // Store the scaled bloom value
        Bloom[CurPixel.xy] = color * colorScale;
    }
}
```

Set the downscaled HDR and average luminance SRVs in that order and the first temporary storage UAV. Then, dispatch the compute shader with the same amount of groups used for the first pass of the luminance downscale.

Next, we are going to filter the values stored in the temporary storage in two passes. Each pass will use the following deceleration for the input and output textures:

```
Texture2D<float4> Input : register( t0 );
RWTexture2D<float4> Output : register( u0 );
```

Both passes will also use the same constant buffer we used for the downscaling and the bright pass. In addition, each pass will be using the following constant array, which contain Gaussian sampling weights (more on that in the *How it works...* section):

```
static const float SampleWeights[13] = {
    0.002216,
    0.008764,
    0.026995,
    0.064759,
    0.120985,
    0.176033,
```

```
        0.199471,
        0.176033,
        0.120985,
        0.064759,
        0.026995,
        0.008764,
        0.002216,
    };


    #define kernelhalf 6
    #define groupthreads 128
```

Finally, both passes will be using the following thread group shared memory deceleration:

```
    groupshared float4 SharedInput[groupthreads];
```

The first pass will filter pixels horizontally from the first temporary texture SRV (containing the bright pass results) into the second temporary texture UAV. The code for the vertical filter is as follows:

```
    [numthreads( groupthreads, 1, 1 )]
    void VerticalFilter( uint3 Gid : SV_GroupID, uint GI : SV_GroupIndex )
    {
        int2 coord = int2( Gid.x, GI - kernelhalf + (groupthreads -
    kernelhalf * 2) * Gid.y );
        coord = clamp( coord, int2(0, 0), int2(Res.x-1, Res.y-1) );
        SharedInput[GI] = Input.Load( int3(coord, 0) );

        GroupMemoryBarrierWithGroupSync();

        // Vertical blur
        if ( GI >= kernelhalf && GI < (groupthreads - kernelhalf) &&
             ( (GI - kernelhalf + (groupthreads - kernelhalf * 2) * Gid.y)
             < Res.y) )
        {
            float4 vOut = 0;

            [unroll]
            for ( int i = -kernelhalf; i <= kernelhalf; ++i )
            {
                vOut += SharedInput[GI + i] * SampleWeights[i +
                kernelhalf];
            }

            Output[coord] = float4(vOut.rgb, 1.0f);
        }
    }
```

Dispatch this compute shader with the same amount of groups in the X dimension as the amount of pixels in the downscaled image width. For the Y dimension, use the following formula: `<down scaled image height> / (128 - 12) + 1`.

Next, we execute a horizontal filter using with the second temporary texture SRV as input and the bloom texture UAV as output. The code for the horizontal filter is as follows:

```
[numthreads( groupthreads, 1, 1 )]
void HorizontalFilter( uint3 Gid : SV_GroupID, uint GI : SV_GroupIndex
)
{
    int2 coord = int2( GI - kernelhalf + (groupthreads - kernelhalf *
2) * Gid.x, Gid.y );
    coord = clamp( coord, int2(0, 0), int2(Res.x-1, Res.y-1) );
    SharedInput[GI] = Input.Load( int3(coord, 0) );

    GroupMemoryBarrierWithGroupSync();

    // Horizontal blur
    if ( GI >= kernelhalf && GI < (groupthreads - kernelhalf) &&
        ( (Gid.x * (groupthreads - 2 * kernelhalf) + GI - kernelhalf)
< Res.x) )
    {
        float4 vOut = 0;

        [unroll]
        for ( int i = -kernelhalf; i <= kernelhalf; ++i )
            vOut += SharedInput[GI + i] * SampleWeights[i +
kernelhalf];

        Output[coord] = float4(vOut.rgb, 1.0f);
    }
}
```

Dispatch this compute shader with the same amount of groups in the Y dimension as the amount of pixels in the downscaled height. For the X dimension, use the following code: `ceil(<down scaled image width> / (128 - 12) )`.

Finally, we need to use the filtered bloom values in the final pass. Add the bloom texture deceleration to the final pass pixel shader along with the linear sampler as follows:

```
Texture2D<float4> BloomTexture : register( t2 );
SamplerState LinearSampler : register( s1 );
```

We will add a bloom scale value at the end of the final pass constant buffer. The updated buffer is declared as follows:

```
cbuffer FinalPassConstants : register( b0 )
{
    // Tone mapping
    float MiddleGrey  : packoffset( c0 );
    float LumWhiteSqr : packoffset( c0.y );
    float BloomScale  : packoffset( c0.z );
}
```

Adding the bloom values to the final image is a single code line change which results in the following pixel shader:

```
float4 FinalPassPS( VS_OUTPUT In ) : SV_TARGET
{
    // Get the color sample
    float3 color = HDRTexture.Sample( PointSampler, In.UV.xy ).xyz;

    // Add the bloom contribution
    color += BloomScale * BloomTexture.Sample( LinearSampler,
    In.UV.xy ).xyz;

    // Tone mapping
    // Find the luminance scale for the current pixel
    float LScale = dot(color, LUM_FACTOR);
    LScale *= MiddleGrey / AvgLum[0];
    LScale = (LScale + LScale * LScale / LumWhiteSqr) / (1.0 + LScale);
    color *= LScale; // Apply the luminance scale to the pixels color

    // Output the LDR value
    return float4(color, 1.0);
}
```

Just set the bloom texture shader resource view as the third texture along with the HDR and average luminance texture's shader resource views.

How it works...

We started the bloom process by downscaling the HDR texture to 1/16 of its size. This is a common practice for different 2D calculations for the following reasons:

- It takes less memory to store these downscaled textures
- Processing downscaled textures are faster because they contain less pixels
- When the downscaled textures is used in the full resolution, the values get an additional cheap blur due to the resolution upscale

This downscaled image can be used in other effects so it's worth keeping it around instead of overwriting it with intermediate values. For that reason we allocated the two temporary textures.

Once we finish the average luminance calculation, the downscaled HDR texture is used in what's called the bright pass. As bloom is a light leakage from very bright pixels, the goal of the bright pass is to find the pixels bright enough to leak into their neighboring pixels. The method used in this recipe to filter the bright pixels is based on luminance values. We want to store a color value higher than black for pixels that are bright enough to be part of the bloom. To do that, we calculate the luminance value for each pixel in the downscaled texture and subtract it from the average luminance scaled by the bloom threshold. Finally, we scale the pixels color with the result. Pixels with a luminance value larger than the scaled average luminance will contribute to the bloom while other pixels will just be black.

With the bright pass results stored in the temporary buffer, it's time to spread them using a Gaussian blur filter in order to achieve the leakage to neighboring pixels. Unlike linear filters that spread values evenly, a Gaussian filter distributes values in a bell shaped curve as shown in the following image:

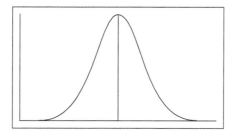

This 2D image shows that an input pixel value gets distributed more towards the center and less for further positions. For a 2D texture, our curve needs to spread values in two dimensions. To implement any blur filter, we use a set of weights which are called the kernel. These weights define how much of the input value will be spread to each output pixel which is within the kernel range. For our Gaussian filter, we want a kernel with a size of 13x13. This means that we would have to write 13x13=169 values for each input pixel. This amount of output pixels will hurt performance, so we use what's called a separable Gaussian blur instead. In this blur technique, we apply the filter in two passes – each pass works on a different dimension. Our first pass will apply a 1D Gaussian filter on the columns while the second pass will do the same for the rows. Each pass will now spread values to only 13 output pixels, which will drastically improve performance. Mathematically, splitting the 2D Gaussian blur into two separate 1D operations is incorrect. Fortunately, it provides a good enough approximation with a much better performance which makes it perfect for this recipe.

To implement the separable Gaussian blur, we store the Gaussian weights in an array of constants as these values never change and can be used for both passes. Each pass starts by loading input pixel values into a thread group shared memory. All threads are synchronized after the load to insure all values are ready. Once the input values are ready, each thread gathers all the weighted input values that are distributed to its corresponding output pixel and stores their sum into the output unordered access view. This technique where a thread gathers its values from shared memory into a single output value is much more efficient compared to having each thread spread a single input value to all the output pixels it gets distributed to.

Once the bloom values are blurred, all we got to do is combine them back with the original HDR image as part of the postprocessing final pass. By using a linear filter, we smoothly upscale the bloom values into the HDR texture resolution. The sampled bloom value is scaled before being added to the HDR color to provide further control on the amount of bloom we want in the final image. Because we add the bloom value to the HDR color before tone mapping is calculated, we can still control the overall brightness of the image which should provide enough tuning options to achieve the final result you are looking for.

Distance depth of field

Depth of field, or **DOF** for short, is a postprocessing technique which emulates lens out of focus. Any lens including the one in our eyes can't focus on the full visible distant range. Up to this point, every pixel in the scene was always rendered in focus regardless of its distance. In this recipe, we will be adding a very simple DOF for far distant objects. This DOF implementation runs very fast and should be used to emulate out of focus towards the end of the visible range.

Getting ready...

We won't need to allocate any new assets for this recipe. For depth sampling, reuse the same depth shader resource view we used for deferred shading. Both projection parameters used to convert ZBuffer depth values to linear range are the same as the ones used for deferred lighting.

How to do it...

All the changes for this recipe will be in the final pass pixel shader. We will be sampling both the downscaled and the depth textures. With those additional textures, the texture deceleration is as follows:

```
Texture2D<float4> HDRTex : register( t0 );
StructuredBuffer<float> AvgLum : register( t1 );
Texture2D<float4> BloomTex : register( t2 );
Texture2D<float4> DOFBlurTex : register( t3 );
Texture2D<float> DepthTex : register( t4 );
```

For the distance DOF calculation, we will need four new constant values. With the new constants, the constant buffer deceleration is as follows:

```
cbuffer FinalPassConstants : register( b0 )
{
    // Tone mapping
    float MiddleGrey : packoffset( c0 );
    float LumWhiteSqr : packoffset( c0.y );
    float BloomScale : packoffset( c0.z );
    float2 ProjectionValues : packoffset( c1 );
    float2 DOFFarValues : packoffset( c1.z );
}
```

All DOF related calculations will be handled by two functions. The first function is the same function we used in deferred lighting to convert ZBuffer depth into linear depth. As a reminder, the code for this calculation is as follows:

```
float ConvertZToLinearDepth(float depth)
{
    float linearDepth = ProjValues.x / (depth + ProjValues.y);
    return linearDepth;
}
```

The second function applies the distance DOF on top of the HDR color. This function is declared as follows:

```
float3 DistanceDOF(float3 colorFocus, float3 colorBlurred, float
depth)
{
    // Find the depth based blur factor
    float blurFactor = saturate((depth - DOFFarValues.x) *
    DOFFarValues.y);

    // Lerp with the blurred color based on the CoC factor
    return lerp(colorFocus, colorBlurred, blurFactor);
}
```

Finally, we need to modify the final pass function to call the preceding function with the input values. The code with the additional function call is as follows:

```
float4 FinalPassPS( VS_OUTPUT In ) : SV_TARGET
{
    // Get the color and depth samples
    float3 color = HDRTex.Sample( PointSampler, In.UV.xy ).xyz;

    // Distance DOF only on pixels that are not on the far plane
    float depth = DepthTex.Sample( PointSampler, In.UV.xy );
```

```
if(depth < 1.0)
{
    // Convert the full resolution depth to linear depth
    depth = ConvertZToLinearDepth(depth);

    // Get the blurred color from the down scaled HDR texture
    float3 colorBlurred = DOFBlurTex.Sample( LinearSampler,
    In.UV.xy ).xyz;


    // Compute the distance DOF color
    color = DistanceDOF(color, colorBlurred, depth);
}

// Add the bloom contribution
color += BloomScale * BloomTex.Sample( LinearSampler,
In.UV.xy ).xyz;

// Tone mapping
// Find the luminance scale for the current pixel
float LScale = dot(color, LUM_FACTOR);
LScale *= MiddleGrey / AvgLum[0];
LScale = (LScale + LScale * LScale / LumWhiteSqr) / (1.0 + LScale);
color *= LScale; // Apply the luminance scale to the pixels color

// Output the LDR value
return float4(color, 1.0);
}
```

How it works...

Similar to the way our eyes use a lens to focus incoming light onto the retina, a camera uses a lens to focus light onto its sensor. DOF is a result of the lens' inability to focus incoming light from sources at different distances at the same time. A term called circle of confusion is used for describing the focal length of a lens. The following image illustrates how incoming light from sources inside the lens' focal length gets properly focused on the sensor:

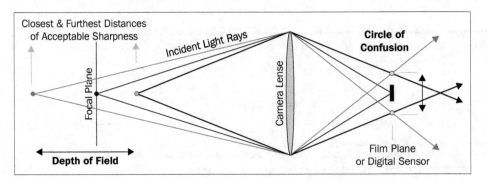

Any incoming light from outside the focal length will show up as a blur in the resulting image. When using this implementation of distance DOF, all pixels close to the camera's near plane end up in focus while pixels far enough from the near plane are out of focus and therefore blurred. We use two new constant values to define the distance from the near plane where our image starts to lose focus and a range after which everything is completely out of focus.

If you look at the changes to the final pass, you will notice that we skip pixels located at the far plane. The reason for that is that we don't want to include areas we didn't render into with this calculation. If you render a scene that fully covers every visible pixel you may remove this condition.

There are other ways to exclude pixels that we don't want to include in the DOF calculation. A very common way is to use the stencil buffer to mark which pixel should be included or excluded from the DOF calculation

Although we can define the out of focus start distance and range with nonlinear depth values, it's easier to tune this effect in world space linear units. For that reason we convert the depth value to linear depth. Converting Z to linear depth inverses the effect of the projection matrix and converts the depth values stored in the ZBuffer back to world space distance from the camera.

Before we can calculate the distance DOF we need to sample the downscaled HDR texture with a linear sampler. By downscaling the HDR texture and then up sampling it we get a cheap blurred version of the HDR texture. We will be using this blurred value to represent the out of focus pixels.

With all three values ready, we call the distance DOF function. First, we calculate how much the pixel is out of focus based on its depth value. The out of focus value has three ranges:

- Pixels with depth closer than the start depth get a value of 0 which is totally in focus.
- Pixels with depth further than the start depth plus range get a value of 1 and are totally out of focus.
- Any other depth value will result with a value between 0 and 1. The higher the value is the more out of focus the pixel is.

To apply the DOF, we linearly interpolate between the HDR sample and the downscaled HDR sample. The closer the out of focus value is to 1 the more we use the blurred downscaled value which makes the pixel look out of focus.

Once all the code changes are in place, you need to tune the start distance and range. Make sure you choose a range larger than 1 to avoid division by zero (DOF won't look good for very low ranges, so you should find a reasonable range). When you set the constant buffer, set 1 over the range we can multiply instead of divide.

Bokeh

Bokeh is another phenomenon related to the lens imperfection. In the world of photography, bokeh describes the way a lens handles out of focus regions. As we already have a way to blur out of focus pixels, we will be focusing on how bokeh handles very bright unfocused pixels. The following image shows how bokeh affects bright light sources that are out of focus:

As you can see, instead of getting blurred, the bright areas in the image form distinct shapes we will call bokeh highlights. This recipe will provide an easy way to render the bokeh highlights.

Getting ready

In this recipe we will accumulate the bokeh shape information into a buffer to later render them on top of the final image. For that we will allocate a new buffer with the following properties:

Buffer descriptor parameter	Value
BindFlags	D3D11_BIND_UNORDERED_ACCESS | D3D11_BIND_SHADER_RESOURCE
StructureByteStride	6 * 4 (6 times size of float)
ByteWidth	128 * StructureByteStride
MiscFlags	D3D11_RESOURCE_MISC_BUFFER_STRUCTURED | D3D11_RESOURCE_MISC_DRAWINDIRECT_ARGS

We will also have a look at the corresponding UAV and SRV for this buffer. For the UAV use the flag D3D11_BUFFER_UAV_FLAG_APPEND so we can append the bokeh shapes as we detect them.

In order to render the content of the bokeh highlights buffer, we need to create a new alpha blending state. Use the following parameters descriptor parameters when creating the blending state:

Blend state descriptor parameter	Value
BlendEnable	True
SrcBlend	D3D11_BLEND_SRC_ALPHA
DestBlend	D3D11_BLEND_INV_SRC_ALPHA
BlendOp	D3D11_BLEND_OP_ADD
SrcBlendAlpha	D3D11_BLEND_ONE
DestBlendAlpha	D3D11_BLEND_ONE
BlendOpAlpha	D3D11_BLEND_OP_ADD
RenderTargetWriteMask	D3D11_COLOR_WRITE_ENABLE_ALL

You will need an alpha texture that represents the shape of the bokeh highlight. Common shape would be a circle or a hexagon which don't have to be uniform. The following image shows an example of how a hexagon can be represented:

As you can see, the shape is completely gray scaled. This is important in order to apply the proper color for each bokeh highlight.

How to do it...

Detecting the pixels bright enough to generate bokeh highlights takes place after the average luminance is calculated. We will be using a compute shader that takes the full resolution HDR, depth, and average luminance as inputs in that order and the new buffer as output. The deceleration for this inputs/output is as follows:

```
Texture2D HDRTex : register( t0 );
Texture2D<float> DepthTex : register( t1 );
StructuredBuffer<float> AvgLum : register( t2 );

struct TBokeh
{
```

```
        float2 Pos;
        float Radius;
        float4 Color;
    };
    AppendStructuredBuffer<TBokeh> Bokeh : register( u1 );
```

In addition, the shader will be using the following constant buffer:

```
    cbuffer BokehConstants : register( b0 )
    {
    uint2 Res : packoffset( c0 );
        float BokehBlurThreshold : packoffset( c0.z );
        float fBloomLumThreshold : packoffset( c0.w );
        float2 ProjValues : packoffset( c1 );
        float2 DOFFarValues : packoffset( c1.z );
        float MiddleGrey : packoffset( c2 );
        float LumWhiteSqr : packoffset( c2.y );
        float RadiusScale : packoffset( c2.z );
        float ColorScale : packoffset( c2.w );
    }
```

The code for the shader is as follows:

```
    [numthreads(1024, 1, 1)]
    void BloomReveal(uint3 dispatchThreadId : SV_DispatchThreadID)
    {
        uint2 CurPixel = uint2(dispatchThreadId.x % Res.x,
    dispatchThreadId.x / Res.x);

        // Skip out of bound pixels
        if(CurPixel.y < Res.y)
        {
            // First we need to find the CoC value for this pixel
            float depth = DepthTex.Sample( PointSampler, In.UV.xy );
            depth = ConvertZToLinearDepth(depth);
            float blurFactor = saturate((depth - DOFFarValues.x) *
            DOFFarValues.y);

            if(blurFactor > BokehBlurThreshold)
            {
                float4 color = HDRTex.Load( int3(CurPixel, 0) );
                float Lum = dot(color, LUM_FACTOR);
                float avgLum = AvgLum[0];

                float colorScale = saturate(Lum - avgLum *
                BokehLumThreshold);
                if(colorScale > 0.0)
                {
```

```
            TBokeh bokeh;
            bokeh.Pos = (2.0 * asfloat(CurPixel) / asfloat(Res)) - 1.0;
            bokeh.Radius = blurFactor * RadiusScale;

            // Tone map the color
            Lum *= MiddleGrey / avgLum;
            LScale = (LScale + LScale * LScale / LumWhiteSqr) /
            (1.0 + LScale);
            bokeh.Color.xyz = HDRColor * LScale * ColorScale;
            bokeh.Color.w = LScale;

            Bokeh.append(bokeh);
          }
        }
      }
    }
```

To execute this shader, call dispatch with the following code: `ceil(<full res HDR texture pixel count> / 1024 )`.

Once the bokeh highlights are accumulated, continue with the same post processing code until the final pass has been completed. We will be rendering the bokeh highlights straight into the LDR target.

We will be using the linear sampler to sample the bokeh highlight shape texture in the pixel shader. In order to alpha blend the bokeh highlights into the LDR texture, set the blend state you prepared in the getting ready section. Set the primitive to `D3D11_PRIMITIVE_TOPOLOGY_POINTLIST` and call `DrawInstancedIndirect` with the pointer to the bokeh highlights buffer to issue the `Draw` call for the highlights.

Use the following deceleration to access the bokeh highlights buffer from the vertex shader:

```
struct TBokeh
{
  float2 Pos;
  float Radius;
  float4 Color;
};
StructuredBuffer<TBokeh> Bokeh : register( t0 );
```

In addition to the bokeh highlights buffer, we will be using the following vertex shader constant buffer:

```
cbuffer BokehDrawConstants : register( b0 )
{
  float RadiusScale : packoffset( c0 );
  float ColorScale : packoffset( c0.y );
}
```

The vertex output structure deceleration is as follows:

```
struct VS_OUTPUT
{
    float4 Pos : SV_Position;
    float Radius : TEXCOORD0;
    float4 Color : COLOR0;
};
```

The code for the vertex shader is as follows:

```
VS_OUTPUT BokehHighlightsVS( uint VertexID : SV_VertexID )
{
    VS_OUTPUT Output;

    TBokeh bokeh = Bokeh[VertexID];
    Output.Position = float4( bokeh.Pos, 0.0, 1.0);
    Output.Radius = bokeh.Radius;
    Output.Color = bokeh.Color;

    return Output;
}
```

Next, we will expend the bokeh highlights into quads using a geometry shader. The geometry shader output structure is as follows:

```
struct GS_OUTPUT
{
    float4 Pos : SV_POSITION;
    float2 UV : TEXCOORD0;
    float4 Color : COLOR0;
};
```

We will be using the following constants where we define each vertex position and UV coordinate (these are the same ones we use in the final pass vertex shader):

```
static const float2 arrBasePos[6] = {
    float2(-1.0, 1.0),
    float2(1.0, 1.0),
    float2(-1.0, -1.0),
    float2(1.0, 1.0),
    float2(1.0, -1.0),
    float2(-1.0, -1.0),
};

static const float2 arrUV[6] = {
    float2(0.0, 0.0),
```

```
            float2(1.0,  0.0),
            float2(0.0,  1.0),
            float2(1.0,  0.0),
            float2(1.0,  1.0),
            float2(0.0,  1.0),
        };
```

The code for the geometry shader is as follows:

```
[maxvertexcount(6)] // Output two triangles per single input point
void BokehGS(point VS_OUTPUT input[1], inout TriangleStream<GS_OUTPUT>
TriStream)
{
 GS_OUTPUT output;

 [unroll]
 for(int i=0, idx = 0; i < 2; i++)
 {
  [unroll]
  for(int j=0; j < 3; j++, idx++)
  {
   // Generate the position
   float2 pos = input[0].Pos.xy + arrBasePos[idx] * input[0].Radius *
   AspectRatio;
   output.Pos = float4(pos, 0.0, 1.0);

   // Copy the UV and color
   output.UV = arrUV[idx];
   output.Color = input[0].Color;

   // Add to the stream
   TriStream.Append(output);
  }

  TriStream.RestartStrip();
 }
}
```

The pixels shader will need the following texture and sampler definition for the bokeh highlight shape:

```
Texture2D<float4> BokehTex : register( t0 );
SamplerState LinearSampler : register( s0 );
```

Finally, the pixel shader code for rendering the bokeh highlights is as follows:

```
float4 BokehHighlightsPS( GS_OUTPUT In ) : SV_TARGET
{
    float4 color = BokehTex.Sample(LinearSampler, In.UV.xy );
    Return color * In.Color;
}
```

How it works...

The idea behind this recipe is to accumulate bokeh highlight instances by scanning the HDR texture for very bright pixels that are out of focus. As we don't know how many highlights we are going to find, the SRV support for thread safe stack like push operator becomes a great way to add the highlights we find during the scan.

In order to find the highlights, we first have to find each pixel CoC blur factor. Only pixels with a blur factor higher than the selected threshold are considered. Considered pixels are then tested to decide if they are bright enough. This test is similar to the one we used for the bright pass. To avoid conflicts with bloom you should choose a higher threshold percentage than the one you did for the bloom bright pass. Each pixel that passed both tests will add a single bokeh highlight to the buffer. As the highlight shape appears larger the blurrier it is, we use the blur factor along with the user defined radius scale to calculate the shape's radius. For the color we have to do the full tone mapping as it is going to get added to the LDR image. Position is stored in clip space and is calculated from the pixel position in the HDR texture.

Our buffer has a limited size which is smaller than the amount of pixels in the HDR image. Once it is filled to its capacity, all further calls to push will have no effect. This behavior means that in case our HDR texture contains too many bright and blurry pixels, the GPU won't crash. When choosing the size of the buffer keep in mind that you probably don't want to support scenes with thousands of highlights.

Rendering the bokeh highlights uses an interesting feature introduced in DirectX10 which lets us issue a `Draw` call without knowing how many vertices or primitives we want to draw. DirectX10 lets us issue this kind of `Draw` call for a vertex buffer that got streamed out and DirectX11 added support for buffers which we use in this recipe. This allows us to issue a `Draw` call with an amount of primitives equal to the amount of highlights we added to the buffer.

The shaders we use to expend the point into a quad are pretty straightforward. The vertex shader reads the bokeh highlight data and passes it to the geometry shader. Inside the geometry shader, we create a quad by adding two triangles to a triangle stream for each bokeh highlight. As the positions are already in clip space they don't have to be transformed. All that is left for the pixel shader is to sample the highlight shape texture and multiply it by the color.

5
Screen Space Effects

In this chapter we will cover various screen space effects:

- ▸ Screen space ambient occlusion
- ▸ Lens flare
- ▸ Screen space Sun rays
- ▸ Screen space reflections

Introduction

In the previous chapter we covered postprocessing effects that enhance the lighting results with a series of 2D calculations in screen space. All the recipes in the previous chapter provide the essentials for enhancing the rendered scene's quality after it is fully lit. Postprocessing, however, is not the only group of calculations that are handled in screen space.

All the recipes covered in this chapter complement our lighting system in some way. As you will soon see, the line between what is considered as postprocessing and what is considered a screen space calculation is not always very clear, but in the end it is not really important as long as you get the resulting image you wanted.

Screen space ambient occlusion

In *Chapter 1, Forward Lighting*, we covered the hemispheric ambient light model where the ambient light is selected based on the vertical component of each pixel normal. You may have noticed that unlike any other light sources we covered, the ambient light does not have a shadow implementation. In fact, our ambient light model doesn't take occlusion into account at all.

In this recipe we will cover a screen space method to calculate the occlusion of each pixel affected by ambient light, and store those occlusion values in what's known as an ambient occlusion map. The ambient occlusion map will later be used during the ambient light calculations to affect the ambient light intensity of each pixel.

Getting ready

All calculations in this recipe are going to be handled at quarter resolution (half the width and half the height of the screen resolution). This is a compromise between performance (less pixels to process is faster) and image quality (smaller image means loss of fine details). We will need one quarter resolution buffer used only by the compute shader and two new textures that will be used to blur the ambient occlusion values. In addition, we will be using two new compute shaders along with a constant buffer, which will be shared by both compute shaders.

Use the following parameters to create the quarter-sized depth and normals buffer:

Buffer descriptor parameter	Value
BindFlags	D3D11_BIND_UNORDERED_ACCESS | D3D11_BIND_SHADER_RESOURCE
StructureByteStride	4 (size of float)
ByteWidth	<Total back buffer pixels>
MiscFlags	D3D11_RESOURCE_MISC_BUFFER_STRUCTURED

All other values in the descriptor should be set to 0.

Once this buffer is allocated, we will need corresponding **unordered access view** (**UAV**) and **shader resource view** (**SRV**), so we can read and write to this buffer. Use the following parameters in the UAV descriptor:

UAV descriptor parameter	Value
Format	DXGI_FORMAT_UNKNOWN
ViewDimension	D3D11_UAV_DIMENSION_BUFFER
Buffer.NumElements	<Total back buffer pixels>/4

All other values in the descriptor should be set to 0.

Use the following parameters in the SRV descriptor:

SRV Descriptor Parameter	Value
Format	DXGI_FORMAT_UNKNOWN
ViewDimension	D3D11_SRV_DIMENSION_BUFFER

All other values in the descriptor should be set to 0.

Next we will need two quarter resolution textures to store the ambient occlusion map values. Use the following parameters to create both textures:

Texture descriptor parameter	Value
Width	\<Back buffer pixel width\> / 2
Height	\<Back buffer pixel height\> / 2
MipLevels	1
ArraySize	1
Format	DXGI_FORMAT_R32_TYPELESS
SampleDesc	1
BindFlags	D3D11_BIND_UNORDERED_ACCESS | D3D11_BIND_SHADER_RESOURCE

All the other values in the descriptor should be set to 0. Once this texture is allocated, we will need a corresponding UAV and SRV, so we can read and write to these textures. Use the following parameters in the UAV descriptor:

UAV descriptor parameter	Value
Format	DXGI_FORMAT _R32_FLOAT
ViewDimension	D3D11_UAV_DIMENSION_TEXTURE2D

All other values in the descriptor should be set to 0. Use the following parameters in the SRV descriptor:

SRV descriptor parameter	Value
Format	DXGI_FORMAT _R32_FLOAT
ViewDimension	D3D11_SRV_DIMENSION_TEXTURE2D

Finally, we allocate the constant buffer with the following parameters:

Constant buffer descriptor parameter	Value
Usage	D3D11_USAGE_DYNAMIC
BindFlags	D3D11_BIND_CONSTANT_BUFFER
CPUAccessFlags	D3D11_CPU_ACCESS_WRITE
ByteWidth	48

All other values in the descriptor should be set to 0.

How to do it...

The ambient occlusion values are computed based on the GBuffer values and are used during the deferred directional light calculation. For this reason, all the steps to calculate the AO values have to take place between the GBuffer and the directional light calculation. We start off by downscaling the depth and normal information we gathered in the GBuffer. Set the depth and normal shader resource views as inputs, and set the depth and normal quarter scale UAV as output. We will be using the following compute shader constant buffer for all computations:

```
cbufferDownScaleConstants : register( b0 )
{
  uint2 Res : packoffset( c0 );
  float2 ResRcp  : packoffset( c0.z );
  float4 ProjParams : packoffset( c1 );
  float4x4 ViewMatrix : packoffset( c2 );
  float OffsetRadius : packoffset( c6 );
  float Radius : packoffset( c6.y );
}
```

Set the following values to the constant buffer:

- `Res`: This is vertical and horizontal quarter-sized resolution
- `ResRcp`: This is the one over the quarter-sized resolution
- `ProjParams`: These are the projection parameters (same ones we used for the deferred lighting)
- `ViewMatrix`: This is the camera's view matrix
- `OffsetRadius`: This is the radius we will use to gather the random positions (see the *How it works...* section for explanation)
- `Radius`: This is the radius of the sphere (see the *How it works...* section for explanation)

Our input SRV and output UAV are defined in the shader as follows:

```
Texture2D<float> DepthTex : register( t0 );
Texture2D<float3> NormalsTex : register( t1 );
RWStructuredBuffer<float4> MiniDepthRW : register( u0 );
```

We will use the following helper function to convert depth buffer values to linear depth:

```
float ConvertZToLinearDepth(float depth)
{
  floatlinearDepth = ProjParams.z / (depth + ProjParams.w);
  returnlinearDepth;
}
```

The following code is the entry point for the downscale compute shader:

```
[numthreads(1024, 1, 1)]
void Downscale(uint3 dispatchThreadId : SV_DispatchThreadID)
{
  uint3 CurPixel = uint3(dispatchThreadId.x % Res.x,
  dispatchThreadId.x / Res.x, 0);

  // Skip out of bound pixels
  if(CurPixel.y<Res.y)
  {
    float minDepth = 1.0;
    float3 avgNormal = float3(0.0, 0.0, 0.0);
    uint3 FullResPixel = CurPixel * 2;

    [unroll]
    for(int i = 0; i< 2; i++)
    {
      [unroll]
      for(int j = 0; j < 2; j++)
      {
        // Get the pixels depth and store the minimum depth
        float curDepth = DepthTex.Load(FullResPixel, int2(j, i));
        minDepth = min(curDepth, minDepth);

        // Sum the viewspacenormals so we can average them
        float3 normalWorld = NormalsTex.Load(FullResPixel, int2(j, i));
        normalWorld = normalize(normalWorld * 2.0 - 1.0);
        avgNormal += mul(normalWorld, (float3x3)ViewMatrix);
      }
    }

    MiniDepthRW[dispatchThreadId.x].x =
    ConvertZToLinearDepth(minDepth);
    //MiniDepthRW[dispatchThreadId.x].yzw = avgNormal * 0.25;

    float3 normalWorld = NormalsTex.Load(FullResPixel, int2(0, 0));
    normalWorld = normalize(normalWorld * 2.0 - 1.0);
    MiniDepthRW[dispatchThreadId.x].yzw = mul(normalWorld,
    (float3x3)ViewMatrix);
  }
}
```

Dispatch the downscale compute shader with amount of X thread groups equal to the total downscale pixels divided by 1,024.

With the depth and normal values stored in the downscaled buffer, we proceed to calculate the AO values. Set the quarter-sized depth normal SRV as an input and the AO UAV as an output. We will be using the following shader deceleration for both of those:

```
Structured Buffer<float4> MiniDepth : register( t0 );
RWTexture2D<float> AO : register( u0 );
```

Our shader will be using the following thread group shared memory:

```
groupshared float SharedDepths[1024];
```

In addition, the shader will be using the following constant values:

```
static const float NumSamplesRcp = 1.0 / 8.0;
static const uint NumSamples = 8;
static const float2 SampleOffsets[NumSamples] = {
  float2(0.2803166, 0.08997212),
  float2(-0.5130632, 0.6877457),
  float2(0.425495, 0.8665376),
  float2(0.8732584, 0.3858971),
  float2(0.0498111, -0.6287371),
  float2(-0.9674183, 0.1236534),
  float2(-0.3788098, -0.09177673),
  float2(0.6985874, -0.5610316),
};
```

We will be using the following code to do the main SSAO calculation:

```
float ComputeAO(int2 cetnerPixelPos, float2 centerClipPos)
{
    // Get the depths for the normal calculation
    float centerDepth = GetDepth(cetnerPixelPos.xy);

    // Find the center pixel veiwspace position
    float3 centerPos;
    centerPos.xy = centerClipPos * ProjParams.xy * centerDepth;
    centerPos.z = centerDepth;

    // Get the view space normal for the center pixel
    float3 centerNormal = GetNormal(cetnerPixelPos.xy);
    centerNormal = normalize(centerNormal);

    // Prepare for random sampling offset
```

```
float rotationAngle = dot(float2(centerClipPos), float2(73.0,
197.0));
float2 randSinCos;
sincos(rotationAngle, randSinCos.x, randSinCos.y);
float2x2 randRotMat = float2x2(randSinCos.y, -randSinCos.x,
randSinCos.x, randSinCos.y);

// Take the samples and calculate the ambient occlusion value for
each
float ao = 0.0;
[unroll]
for(uint i=0; i < NumSamples; i++)
{
  // Find the texture space position and depth
  float2 sampleOff = OffsetRadius.xx * mul(SampleOffsets[i],
  randRotMat);
  float curDepth = GetDepth(cetnerPixelPos + int2(sampleOff.x,
  -sampleOff.y));

  // Calculate the view space position
  float3 curPos;
  curPos.xy = (centerClipPos + 2.0 * sampleOff * ResRcp) *
  ProjParams.xy * curDepth;
  curPos.z = curDepth;
  float3 centerToCurPos = curPos - centerPos;
  float lenCenterToCurPos = length(centerToCurPos);
  float angleFactor = 1.0 - dot(centerToCurPos / lenCenterToCurPos,
  centerNormal);
  floatdistFactor = lenCenterToCurPos / Radius;

  // Sum the samples AO factors
  ao += saturate(max(distFactor, angleFactor));
}

  return ao * NumSamplesRcp;
}
```

The entry code for the SSAO shader is as follows:

```
[numthreads(1024, 1, 1)]
void SSAOCompute(uint3 groupThreadId : SV_GroupThreadID, uint3
dispatchThreadId : SV_DispatchThreadID)
{
  uint2 CurPixel = uint2(dispatchThreadId.x % Res.x,
  dispatchThreadId.x / Res.x);

  SharedDepths[groupThreadId.x] = MiniDepth[dispatchThreadId.x].x;
```

```
GroupMemoryBarrierWithGroupSync();

// Skip out of bound pixels
if(CurPixel.y<Res.y)
{
    // Find the XY clip space position for the current pixel
    // Y has to be inverted
    float2 centerClipPos = 2.0 * float2(CurPixel) * ResRcp;
    centerClipPos = float2(centerClipPos.x - 1.0, 1.0 -
    centerClipPos.y);

    AO[CurPixel] = ComputeAO(CurPixel, centerClipPos);
}
}
```

Dispatch the SSAO compute shader with amount of X thread groups equal to the total downscale pixels divided by 1,024.

At this point, we get the AO values computed and we need to blur them. We will be using the exact same blur filter we used for the depth of field downscaled texture (see the *Distance depth of field* recipe of *Chapter 4, Postprocessing*, for the implementation). Once the blur step is completed, we can use the blurred AO values during the directional light calculation. Set the blurred AO SRV along with the other GBuffer SRV we normally use for directional lighting. Add the following texture deceleration to the directional light pixel shader:

```
Texture2D<float>AOTex: register( t4 );
```

Replace the pixel shader entry point with the following code:

```
float4 DirLightPS(VS_OUTPUT In) : SV_TARGET
{
    // Unpack the GBuffer
    float2 uv = In.Position.xy;//In.UV.xy;
    SURFACE_DATA gbd = UnpackGBuffer_Loc(int3(uv, 0));

    // Convert the data into the material structure
    Material mat;
    MaterialFromGBuffer(gbd, mat);

    // Reconstruct the world position
    float2 cpPos = In.UV.xy * float2(2.0, -2.0) - float2(1.0, -1.0);
    float3 position = CalcWorldPos(cpPos, gbd.LinearDepth);

    // Get the AO value
    float ao = AOTex.Sample( LinearSampler, uv.xy );
```

```
// Calculate the light contribution
float4 finalColor;
finalColor.xyz = CalcAmbient(mat.normal, mat.diffuseColor.xyz) * ao;
finalColor.xyz += CalcDirectional(position, mat);
finalColor.w = 1.0;

returnfinalColor;
}
```

Now the directional light takes the AO values when calculating the ambient light.

How it works...

As you may recall from *Chapter 1*, *Forward Lighting*, ambient light has no origin; it affects pixels based only on the pixels' normal direction. Without a light origin and direction, using a shadow map to represent occlusion is no longer trivial. Ambient occlusion provides an alternative way to attenuate light influence based on the scene's geometry occlusion that works well for ambient lights.

Unlike shadow maps, which represent the depths of the light occluders from the light's point of view, ambient occlusion provides an attenuation value for each visible pixel based on the amount of occlusion generated by its surrounding. The following figure shows an example of a given pixel's occlusion (the area inside the blue arch):

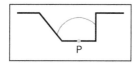

As you can see, the occlusion of the given pixel is defined by the angle of the widest cone that doesn't intersect with the rest of the scene. The wider the cone's angle is, the less occluded the pixel is and the more ambient light it will receive. The cone angle for a given pixel may dramatically change based on the cone's distance; the longer the cone, the more chance it has to intersect with the scene that will increase the pixel's occlusion value. An ambient occlusion map stores a value that indicates the amount of occlusion for each pixel, where a value 1 represents a nonoccluded pixel and a value 0 represents a fully occluded pixel.

Now that we have defined what the ambient occlusion map is and what it contains, we can go over the method we use to calculate the ambient occlusion values. The technique we are using is called **screen space ambient occlusion. (SSAO)**. When this technique was first introduced, the common way to calculate SSAO value for a given pixel was to compare random positions inside a sphere centered at that pixel position, and compare each position's depth with the depth stored in the depth buffer. Positions with a depth value further than the one stored in the depth buffer were considered to be occluded. The final AO value of each pixel was the total of nonoccluded samples divided by the total samples. The following illustration demonstrates how this method works with the test sphere colored blue and random positions colored gray:

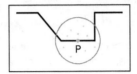

As you can see, this SSAO implementation is not very efficient, as some of the samples will always end up occluded as they are positioned behind the center pixel. In order to avoid this waste, the solution we are using avoids using explicit positions. Instead, our SSAO implementation reconstructs eight random view space positions around the pixel using the depth buffer (same method used when calculating positions for deferred shading). Each position is tested against the sphere to see if it's inside our cone range. In addition, we test the angle between the pixel's normal and the vector connecting the pixel and the random position. The AO value for each random position is closer to the wider angle or to the position that is further from the sphere center. As seen before with the original implementation, the final AO value for the pixel is the normalized sum of the random position AO value. The following figure demonstrates how the test for one of the positions will look like (note that the AO value for position $p\~$ is going to be close to 1 as the angle is close to 90 degrees and the position is close to the edge of the sphere):

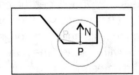

As mentioned in the *Getting ready* section, we calculate SSAO in the quarter scale of the final image. Downscaling the normal and depth values is straight forward. For depth values we convert the depth buffer values to linear depth, find the minimum value (the value closest to the camera) of every group of 4 pixels, and store that value into the corresponding quarter-sized buffer slot. Normal values are retrieved from the GBuffer, transformed from world space to view space, averaged, and then stored into the corresponding quarter-sized buffer slot.

Once the values have been downscaled, we dispatch the SSAO compute shader to handle the actual AO value calculation. We load the depth and normal for the pixel we process and reconstruct its view space position. We then execute a loop that handles the random position AO calculation. We will be using a fixed set of eight offsets defined in `SampleOffsets` to sample the downscaled depth with. To break the pattern of the fixed offsets, we come up with an angle based on the pixel position that is used to construct a 2D rotation matrix. Constructing this rotation matrix is described in the following illustration:

The fixed offsets are then rotated with that matrix and multiplied by a fixed sample radius value. Using the rotated offset, we sample the downscaled depth buffer and construct a view space position for that sample. We then proceed to calculate the vector connecting the pixel and our sample view space position. Once we have the vector between the pixel and the sample, we calculate two values, distance factor and angle factor, which will be used to determine the final AO value.

The distance factor is the vector's length divided by the sphere radius. This factor reduces the occlusion, that is, the further the sample is from the pixel. This insures that samples that are outside the sphere radius are considered fully visible.

The angle factor is based on the dot factor between the vector and the pixel's normal. We consider an angle range of 0 to 90 as a transition from fully occluded to not occluded at all. Every angle above 90 is also considered as not occluded at all. The angle factor is calculated as 1 minus the dot factor between the vector and the normal.

Once we have these two factors, we use the saturated maximum of the two values (the more exposed value) as the sample's AO value. The AO values are summed and normalized, and stored in the buffer. Once the compute shader is done, the resulting AO map is very grainy due to the sample rotation. We need to blur it in order to smooth those values. This operation was already covered in the *Bloom* recipe of *Chapter 4*, *Postprocessing*.

With the AO map properly blurred, the only thing left is to use it during the ambient light calculation. As you may recall, we calculate the ambient light as part of the directional light, so we will sample the AO map in that shader. We use linear sampling for the AO map to smooth it a bit further. Using the AO map in the pixel shader is very simple; for each pixel the AO value is a scale we can apply directly to the ambient light value by multiplying the two.

Lens flare

As mentioned in some of the recipes of *Chapter 4, Postprocessing*, some of the techniques we covered copy qualities common to a camera lens instead of a human eye. Lens flare is yet another camera lens imperfection commonly seen in movies. Usually seen when a very bright light source is present (such as the Sun), lens flare shows up as a series of colorful shapes that cover the scene. As an example, the following screenshot showing the Sun close to the top-right corner and a series for lens flares spread over the screenshot's diagonal:

For this recipe we are going to generate lens flares only for the Sun. This implementation can be easily modified to support other light sources. See the *There's more...* section for additional details.

Getting ready

For this example we will be adding lens flares only for the Sun. See the *There's more...* section for details about extending the support to other light sources.

Before we start rendering the lens flares, we will need to decide if the light source causing the flares is visible or not. In addition, we will have to get an estimate of how much of the light source is actually visible. We will need to create two types of D3D queries to handle this: one is an occlusion query and the other is a predicate query. Create the occlusion query by calling `CreateQuery` with `D3D11_QUERY_OCCLUSION` in the query descriptor. Create the predicate by calling `CreatePredicate` with `D3D11_QUERY_OCCLUSION_PREDICATE` in the query descriptor.

We will be using two textures to represent the different elements of the flare. First texture will be used for a flare called the aperture and the other will be used to assemble the corona. You should load those two textures into two shader resource views. The following screenshot is an example of what an aperture texture may contain:

An example for a corona texture is presented in the following screenshot:

Rendering the flares will require an additive blending and depth-disabled render states. Since we already defined the additive blend state for lighting and the depth disabled state for postprocessing, we can just reuse those states while rendering the Sun's lens flares.

Some of the Sun lens flares will be positioned along a line passing through the Sun position and the camera's center. The further the Sun position is from the center of the screen, the more spread the flares are across this line. To calculate the flare's positions, we will need to transform the Sun's position from world space to screen space before we render the lens flares. To find the Sun's screen space position, just multiply the Sun's position by the camera's world view projection matrix and divide the resulting vector's x and y components by the w component. To find the vector used for positioning the lens flares, subtract a zero vector from the Sun's 2D screen space position.

We will be rendering multiple flares whenever the Sun is visible. Each flare we will draw requires a fixed set of parameters that doesn't change. These parameters will control the position and look of each flare, so their values should be tuned to match the visual result that is most suitable for you. The first parameter is an offset that will determine the flares position along the vector connecting the screen center and the Sun position. In addition to the offset, the flares will also require a scale and rotation values. We will be using a 2D matrix to represent both, rotation and scaling. Calculating the scale rotation matrix was covered in the *Screen space ambient occlusion* recipe.

The following figure is a reminder:

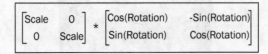

Finally, each flare has its own color.

How to do it...

Rendering the Sun's lens flares is handled in two stages: issuing the Sun visibility queries and the actual rendering. The queries are handled when the Sun is rendered. Use the immediate context's begin and end with the predicate query around the Sun draw. Note that if you skip rendering the Sun for any reason, the predicate value will be wrong and you will have to skip rendering the flares by other means. In addition to the predicate, we will need to issue the occlusion query to determine how many pixels got rendered for the Sun. Unlike the predicate, the occlusion query can't be issued every frame, because we want to read back the results on the CPU. To handle this correctly, you will need to issue the occlusion query only after the previous results were ready.

Once we are done with the queries, we can proceed to the second stage and render the flares. Make sure that the flares are the last thing getting rendered before the postprocessing takes place, so they don't get overwritten by other draw calls. Before rendering the flares, you will need to calculate their screen space position and intensity (see the *How it works...* section for more details about the two different flare types and their corresponding values). The screen space position of each flare is calculated by multiplying the offset by the vector we found in the *Getting ready* section. The intensity of the each flare has to be scaled based on the results of the occlusion query. You will need to find the amount of pixels rendered when the Sun is fully visible once, so we can normalize the occlusion query result. To find the intensity scale, divide the value returned by the occlusion query by that maximum visibility value and multiply each flare's intensity by this scale before passing it to the shader.

We will render the Sun's lens flares with two draw calls, each with a different texture. Use D3D_PRIMITIVE_TOPOLOGY_TRIANGLELIST as the primitive topology, and set the input layout and vertex buffer to null in the same way we used it for the deferred directional light.

Our vertex shader will be using a constant buffer with arrays of values for each one of the flares getting rendered. This constant buffer will contain an array with the values of each one of the flare's. Use the following deceleration for the constant buffer and its internal per-flare structure:

```
static constint MAX_FLARES = 8;

struct FLARE
{
```

```
   float4 Position;
   float4 ScaleRotate;
   float4 Color;
};

cbuffercb Lensflare : register( b0 )
{
   FLARE Flares[MAX_FLARES] : packoffset( c0 );
}
```

Note that this storage wastes the Z and W components of the position as we use screen space 2D positions. This unused memory can be used for additional parameters you may want to add to extend your control on the flares final look.

Depending on whether you use DirectX 11 or DirectX 11.1, the value of MAX_FLARES is either the maximum amount of aperture flares we would like to render (DirectX 11) or the maximum amount of aperture flares plus three for the corona (DirectX 11.1). On DirectX 11 you will need to fill the buffer separately for each draw call (or use two separate buffers). Start with the corona values and add the aperture values afterwards. On DirectX 11.1 you can fill the buffer once with all the values, and set the suitable range for each one of the draw calls. Again, we will need the corona values first and the aperture ones afterwards.

Once the buffer is ready, set the corona texture view to the first slot and call Draw with 18 vertices. Set the aperture texture to slot one and call Draw with six times the amount of aperture flares as the amount of vertices.

The vertex shader we will be using will output the following structure:

```
struct VS_OUTPUT
{
   float4 Pos : SV_Position;
   float2 UV : TEXCOORD0;
   float4 Col : TEXCOORD1;
};
```

Similar to the way we generated the quad for the deferred directional light, each six vertices will form a pair of triangles positioned into a quad. We will be using the following shader constants for the position and UV values of each vertex:

```
static const float2 arrBasePos[6] = {
   float2(-1.0, 1.0),
   float2(1.0, 1.0),
   float2(1.0, -1.0),
   float2(-1.0, 1.0),
   float2(1.0, -1.0),
```

```
        float2(-1.0, -1.0),
    };

    static const float2 arrUV[6] = {
        float2(0.0, 0.0),
        float2(1.0, 0.0),
        float2(0.0, 1.0),
        float2(1.0, 0.0),
        float2(1.0, 1.0),
        float2(0.0, 1.0),
    };
```

The vertex shader code is as follows:

```
    VS_OUTPUT LensflareVS(uint VertexID : SV_VertexID )
    {
        VS_OUTPUT Out;

        // Get the flares values
        FLARE flare = Flares[VertexID / 6];

        // Calculate the clip space position
        float2 pos2D;
        pos2D.x = dot(arrBasePos[VertexID % 6].xy, flare.ScaleRotate.xy);
        pos2D.y = dot(arrBasePos[VertexID % 6].xy, flare.ScaleRotate.zw);
        pos2D += flare.Position.xy;
        Out.Pos = float4(pos2D, 0.0, 1.0);

        // Pass the flare color
        Out.Col = flare.Color;

        // Pass the vertex UV
        Out.UV = arrUV[VertexID % 6];

        return Out;
    }
```

This vertex shader outputs all the information needed for the pixel shader, so there is no need for a pixel shader constant buffer. We will be using the following texture and sampler for the flare's texture dictionary:

```
    Texture2D<float4> FlareTexture : register( t0 );
    SamplerState LinearSampler : register( s0 );
```

Here is the code for the pixel shader:

```
float4 LensflarePS( VS_OUTPUT In ) : SV_TARGET
{
   float4 texColor = FlareTexture.Sample(LinearSampler, In.UV);

   // Convert to linear color space
   texColor.rgb *= texColor.rgb;

   // Scale by the flares color
   return texColor * In.Col;
}
```

How it works...

A camera lens is usually assembled from multiple lenses with different shapes and sizes. This array of lenses focuses the incoming light rays with the camera's sensor or film. In a perfect world, all the light entering the lens array will go through each lens, eventually exit the final lens, and hit the sensor. In reality, some of the light rays hitting the lens may get reflected instead of passing through. Lens flares are the result of rays getting reflected from a lens that eventually find their way to the sensor, only they are not focused the way they should due to the undesired trip they undertook. Even though this means that lens flares are an undesired artifact, photographers and film makers learned how to take advantage of lens flares to their artistic needs that make them commonly visible.

Depending on the degree of realism you hope to achieve, lens flares can be either physically simulated or estimated with a few simple calculations. Games usually can't afford (and don't need) to spend too much time simulating this effect, so they go with the estimated implementation. Using an estimate is not a big compromise, as this effect looks pretty convincing even with this simplified implementation.

Each camera lens generates a different amount of flares with different shapes. The shapes are mostly dictated by the aperture shape while the amount depends on the complexity of the lens array. Probably the most common flare shape is an octagon generated from the camera's aperture. These types of flares change position based on the vector we used in the *Getting ready* section, so the offset they use should be greater than zero. Another common flare shape is a set of rays emitted from the light source center position called the corona. The corona is always positioned at the light source center, which means it has an offset of zero. If you want to increase the visual richness of the lens flare effect, just add additional shapes and increase the amount flares.

Before the flares can be rendered, we need to calculate the values we will pass to the constant buffer. Each flare's color should be scaled by the Sun's visibility factor. Make sure that the color is in linear space before scaling it. The corona should rotate slowly for as long as it's visible, while the aperture flares should normally use a fixed rotation.

Rendering the flares is handled in two separate draw calls: the corona's rendered first using three quads and the aperture flares are rendered afterwards with one quad per flare. In the vertex shader, we use the vertex ID to choose a different set of positions and UV coordinates for each vertex similar to how we handled quads in previous recipes. Each vertex position is then transformed by the scale and rotation values.

The pixel shader samples the texture and squares the value to convert it to a linear color space. The linear texture sample is then multiplied by the input color to get the final pixel color. Since the output color is an HDR value, this effect has to be rendered before postprocessing. Since the flares tend to be bright, the bloom effect is going to add a nice touch to the flares in the final image.

There is more...

Lens flares are not restricted just to the Sun, but with a little effort you can support lens flares for multiple light sources in the scene. One thing to take into account is that DirectX has a limitation on the amount of predicates and occlusion queries that can be created. A good strategy would be to decide how many light sources can be supported at the same time. During runtime you can gather the most dominant lights every frame and render the flares only for those lights.

Light sources other than the Sun will usually be less intense. You can save some performance by rendering a simpler set of aperture flares for those light sources and scale their flares to a smaller size.

Screen space reflections

It is very likely that at one point you will want to render a mesh with reflective qualities. Rendering reflection in 3D is usually a performance-costly task that is limited to planer reflections or environment reflection. Another common reflection restriction is that a scene cannot contain more than one reflective surface at any given time. Both restrictions usually provide an annoying challenge for artists, as they have to take these limitations into account when choosing the materials used in the scene.

Screen space reflection provides an alternative way to support reflection without the previously mentioned restrictions. This technique can be used for any surface shape and it is not restricted to a single surface in a given frame. Although screen space reflections are limited by the lack of information available in screen space, they offer a good alternative in scenes that can't use the traditional reflection methods, as you can see in the following example:

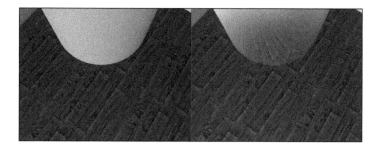

Getting ready

For this recipe we will need to sample the reflection values from the light accumulation buffer and write the result back into the same buffer. Since we are not allowed to read and write to the same buffer, we will need a second HDR texture to write the reflection values into while reading from the light accumulation texture. At the end of this phase, the new reflection HDR texture will get combined with the light accumulation texture. When allocating the reflections HDR texture and its corresponding render target view and SRV, use the same parameters we used for the main HDR light accumulation texture and views.

Blending the reflection HDR texture with the light accumulation HDR texture is going to be implemented using an additive blend state. We will be adding the reflection color on top of the light colors based on the reflection color opacity. In order to create this alpha blend state, use the following values in its descriptor:

Blend state descriptor parameter	Value
BlendEnable	TRUE
SrcBlend	D3D11_BLEND_SRC_ALPHA
DestBlend	D3D11_BLEND_ONE
BlendOp	D3D11_BLEND_OP_ADD
SrcBlendAlpha	D3D11_BLEND_SRC_ALPHA
DestBlendAlpha	D3D11_BLEND_INV_SRC_ALPHA
BlendOpAlpha	D3D11_BLEND_OP_ADD
RenderTargetWriteMask	D3D11_COLOR_WRITE_ENABLE_ALL

In addition to the new reflection HDR texture and its blend state, we will be sampling the depth values. To allow depth tests while sampling from the depth buffer, we will use the same read-only depth shader view with depth's writes-disabled solution similar to previous recipes.

How to do it...

For this recipe we assume that all the reflective surfaces we will use are fully opaque, so the screen space reflections are calculated right after we are done lighting the GBuffer and rendering any emitting meshes. Set the depth state to the "no depth writes" state with the read-only depth stencil view. Set the new HDR render target view and clear it to fully black with a zero alpha. Set the depth SRV and the light accumulation SRV to the first and second slots of the pixel shader.

We will be rendering each mesh that uses the screen space reflection with the same vertex and pixel shaders with the option to change the constant buffer values for each mesh based on its properties. The vertex shader constant buffer has to contain all the information required to transform the mesh into projected space. The minimum needed buffer is as follows:

```
cbuffer SSReflectionPSConstants : register( b0 )
{
  float4x4 WorldViewProjection : packoffset( c0 );
  float4x4 WorldView : packoffset( c4 );
}
```

The vertex shader we will use outputs the following structure:

```
struct VS_OUTPUT
{
  float4 Position : SV_Position;
  float4 ViewPosition : TEXCOORD0;
  float3 ViewNormal : TEXCOORD1;
  float3 csPos : TEXCOORD2;
};
```

The code for the vertex shader is as follows:

```
VS_OUTPUT SSReflectionVS(float4 pos : POSITION, float3 norm : NORMAL)
{
  VS_OUTPUT Output;

  // Transform to projected space
  Output.Position = mul(pos, WorldViewProjection);

  // Transform the position and normal to view space
  Output.ViewPosition = mul(pos, WorldView);
  Output.ViewNormal = mul(norm, (float3x3)WorldView);

  // Convert the projected position to clip-space
  Output.csPos = Output.Position.xyz / Output.Position.w;

  return Output;
}
```

Again, this is the minimum vertex shader needed for the screen space reflection. If the mesh getting rendered uses a more sophisticated vertex shader, a geometry shader or a tessellation stage, the same set of shaders is required for the screen space reflections.

For our pixel shader, we will use the following constant buffer:

```
cbuffer SSReflectionVSConstants : register( b0 )
{
  float4x4 ProjMatrix : packoffset( c0 );
  float ViewAngleThreshold : packoffset( c4 );
  float EdgeDistThreshold : packoffset( c4.y );
  float DepthBias : packoffset( c4.z );
  float ReflectionScale : packoffset( c4.w );
}
```

We will be using a linear sampler to sample the HDR color, depth, and normal textures that are declared in the pixel shader as follows:

```
Texture2D<float4> HDRTex : register( t0 );
Texture2D<float> DepthTex : register( t1 );
Texture2D<float3> NormalTex : register( t2 );

SamplerState PointSampler : register( s0 );
```

Linear depth conversion is handled as usual by the following helper function:

```
float ConvertZToLinearDepth(float depth)
{
  floatlinearDepth = ProjParams.z / (depth + ProjParams.w);
  return linearDepth;
}
```

We will be using the following constant values in the pixels shader:

```
static const float PixelSize = 2.0 / 768.0f;
static const int nNumSteps = 1024;
```

Both values are resolution dependent (in this case 1024x768), so change this values to match your light accumulation buffer size.

Finally, the pixel shader code is as follows:

```
float4 SSReflectionPS( VS_OUTPUT In ) : SV_TARGET
{
  // Pixel position and normal in view space
  float3 vsPos = In.ViewPosition.xyz;
  float3 vsNorm = normalize(In.ViewNormal);
```

```
// Calculate the camera to pixel direction
float3 eyeToPixel = normalize(vsPos);

// Calculate the reflected view direction
float3 vsReflect = reflect(eyeToPixel,  vsNorm);

// The initial reflection color for the pixel
float4 reflectColor = float4(0.0, 0.0, 0.0, 0.0);

// Don't bother with reflected vector above the threshold vector
if (vsReflect.z > ViewAngleThreshold)
{
 // Fade the reflection as the view angles gets close to the threshold
 float viewAngleThresholdInv = 1.0 - ViewAngleThreshold;
 float viewAngleFade = (vsReflect.z - ViewAngleThreshold) /
 viewAngleThresholdInv;

 // Transform the View Space Reflection to clip-space
 float3 vsPosReflect = vsPos + vsReflect;
 float3 csPosReflect = mul(float4(vsPosReflect, 1.0), ProjMatrix).xyz
 / vsPosReflect.z;
 float3 csReflect = csPosReflect - In.csPos;

 // Resize Screen Space Reflection to an appropriate length.
 float reflectScale = PixelSize / length(csReflect.xy);
 csReflect *= reflectScale;

 // Calculate the offsets in texture space
 float3 currOffset = In.csPos + csReflect;
 currOffset.xy = currOffset.xy * float2(0.5, -0.5) + 0.5;
 float3 lastOffset = In.csPos;
 lastOffset.xy = lastOffset.xy * float2(0.5, -0.5) + 0.5;
 csReflect = float3(csReflect.x * 0.5, csReflect.y * -0.5,
 csReflect.z);

 // Iterate over the HDR texture searching for intersection
 for (int nCurStep = 0; nCurStep < nNumSteps; nCurStep++)
 {
  // Sample from depth buffer
  float curSample = DepthTex.SampleLevel(PointSampler,
  currOffset.xy, 0.0).x + DepthBias;
  if (curSample < currOffset.z)
  {
```

```
    // Correct the offset based on the sampled depth
    currOffset.xy = lastOffset.xy + (currOffset.z - curSample) *
    csReflect.xy;

    // Get the HDR value at the location
    reflectColor.xyz = HDRTex.SampleLevel(PointSampler,
    currOffset.xy, 0.0).xyz;

    // Fade out samples close to the texture edges
    float edgeFade = saturate(distance(currOffset.xy,
    float2(0.5, 0.5)) * 2.0 - EdgeDistThreshold);

    // Find the fade value
    reflectColor.w = min(viewAngleFade, 1.0 - edgeFade * edgeFade);

    // Apply the reflection sacle
    reflectColor.w *= ReflectionScale;

    // Advance past the final iteration to break the loop
    nCurStep = nNumSteps;
   }

   // Advance to the next sample
   lastOffset = currOffset;
   currOffset += csReflect;
  }
 }

 return reflectColor;
}
```

Once all the meshes with reflective material are rendered, our reflection HDR render target contains the reflection values for all the pixels that got rendered using a reflective material. In order to combine it with the light accumulation HDR texture, we will need to use alpha blending.

Set back the render target view of the light accumulation texture, so we can blend the reflections into it (no need for a depth target), set the blend state to the additive blend state, and set the reflection HDR texture SRV into the first pixel shader resource slot.

We will render a full screen quad similar to the way we did in previous recipes. Our vertex shader will be using the following global constants for the full screen vertex positions:

```
static const float2 arrBasePos[4] = {
    float2(1.0,  1.0),
    float2(1.0,  -1.0),
    float2(-1.0, 1.0),
    float2(-1.0, -1.0),
};
```

The code for the vertex shader is as follows:

```
loat4 ReflectionBlendVS( uint VertexID : SV_VertexID ) : SV_Position
{
    // Return the quad position
    return float4( arrBasePos[VertexID].xy, 0.0, 1.0);
}
```

Our pixel will sample from the reflection texture SRV using the following texture and sampler deceleration:

```
Texture2D<float4> HDRTex : register( t0 );

SamplerState PointSampler : register( s0 );
```

Finally, the following code is used for the pixel shader:

```
float4 ReflectionBlendPS(VS_OUTPUT In) : SV_TARGET
{
    return HDRTex.Load(int3(In.Position.xy, 0));
}
```

To draw the full screen quad, set the topology to `D3D_PRIMITIVE_TOPOLOGY_TRIANGLESTRIP`, set the vertex and index buffers to `NULL`, and call a draw with four vertices. Make sure that you restore the previous states after this draw call.

How it works...

Calculating reflections in screen space is implemented using ray tracing in texture space. By gathering samples along the reflection vector, we can find the first intersection between the reflected ray and the scene. The color stored in the light accumulation buffer in the ray intersection position is scaled and stored as the reflected color. The main problem with this technique is that we are missing the information outside the visible area. Even with this limitation, screen space reflections provide pleasing results when tuned correctly.

Executing the pixel shader for the pixels that use a reflective material can be handled in two ways: first option is to render a full screen quad and skip pixels that are not reflective using stencil test and the second option is to render a mesh that represents the reflective surface that will affect only the reflective pixels. This recipe uses the second option to avoid the need to tag the reflective pixels using stencil. Note that using stencil is probably faster, because it can also be used to skip pixels on the second draw call that blends the reflections with the light accumulation texture. Another advantage of the first option is that you no longer need to execute the proper vertex shader for meshes that are skinned or use tessellation, all you need is to draw a full screen quad.

In the pixel shader, we handle the ray tracing in two stages. First, we gather all the data we will need and then we execute a loop that checks for intersections along the ray. In order to do the ray tracing, we get the depth and normal from the textures. The depth is converted to linear depth, and the normal is normalized and converted to view space. We then proceed to calculate the reflected opacity based on its angle.

Since we don't have the information behind the camera, we can't find the reflection color for any pixel with a normal pointing towards the camera. To go around this problem, we increase the opacity based on the angle between the normal and the camera direction with an addition passed as `ViewAngleThreshold`. In addition, we set the opacity to fully transparent for any angle wider than 90 degrees.

In order to ray trace in texture space, we need a reflection vector. We calculate the reflection vector using the HLSL reflect function. The resulting reflection vector is in view space, so we have to transform it to clip space using the projection parameters. We then prepare the sampling offset along the clip space vector.

At this point, we have completed the first stage and we can start the ray tracing. We march along the clip space projection ray using eight iterations (you may change this value based on quality and performance decisions). Each iteration of the `for` loop samples depth and color values along the clip space reflection vector with a larger distance compared to the previous one as illustrated in the following figure:

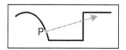

For a position `P`, we sample the depth along the blue vector with each sample that doesn't hit anything marked with a blue line and the first hit marked with a red line. Hits are detected by comparing the absolute difference between the sample view position linear depth and the corresponding depth buffer linear depth. If the deference is smaller than `DepthRange`, we consider this sample a hit and will ignore the rest of the samples through `isSampleFound`.

In order to reduce the chance of artifacts, we reduce the intensity of the reflection color (if found) based on its distance along the ray; the further it is from the pixel position, the lower its alpha value is going to be.

Once all iterations are completed, the final alpha value is calculated and the pixel is written to the reflection HDR texture. With all the values written into the reflection HDR texture, we alpha blend them with the light accumulation values using additive blending based on the alpha values stored in the reflection map.

Screen space Sun rays

Sun rays are bright light streaks that normally show up when the Sun is partially or completely occluded and rays of light stretch from the edges of the occlusion outwards. It is very common to see this phenomenon when thick clouds obscure the Sun and the rays are visible through the gaps between those clouds. This effect's scientific name is Crepuscular rays, but it goes by many other names such as God rays, Sun rays, and light shafts. In games, photography, and movies, it is not uncommon to see Sun rays either generate from a bright light source other than the Sun or due to occlusion from objects other than clouds. The following screenshot demonstrates how Sun rays can be seen when the Sun is occluded by an opaque object:

There are multiple ways to calculate light rays, each with different limitations and performance requirements. In this recipe, we will approximate the Sun light rays using a screen space calculation. As with all the other screen space techniques, we will refer to this technique as SSLR.

Getting ready

The light rays require two quarter resolution 2D textures along with corresponding SRVs, UAVs, and RTV. Use the following parameters to create both textures:

Texture descriptor parameter	Value
Width	<Back buffer pixel width> / 2
Height	<Back buffer pixel height> / 2
MipLevels	1
ArraySize	1
Format	DXGI_FORMAT_R8_TYPELESS
SampleDesc	1
BindFlags	D3D11_BIND_UNORDERED_ACCESS | D3D11_BIND_SHADER_RESOURCE

All other values in the descriptor should be set to 0.

Once this texture is allocated, we will need a corresponding **shader resource view** (**SRV**) for both textures, so we can sample them. Use the following parameters in the SRV descriptor:

SRV Descriptor Parameter	Value
Format	DXGI_FORMAT _ R8_UNORM
ViewDimension	D3D11_SRV_DIMENSION_TEXTURE2D

All other values in the descriptor should be set to 0.

Only one of the textures requires UAV. Use the following parameters in the UAV descriptor:

UAV Descriptor Parameter	Value
Format	DXGI_FORMAT _R8_UNORM
ViewDimension	D3D11_UAV_DIMENSION_TEXTURE2D

All other values in the descriptor should be set to 0.

Both textures require a **render target view** (**RTV**). Use the following parameters in the RTV descriptor:

RTV Descriptor Parameter	Value
Format	DXGI_FORMAT _R8_UNORM
ViewDimension	D3D11_RTV_DIMENSION_TEXTURE2D

All other values in the descriptor should be set to 0.

We will be using two additive blend states. Use the following parameters for the blend state descriptor:

Blend State Descriptor Parameter	Value
BlendEnable	TRUE
SrcBlend	D3D11_BLEND_ONE
DestBlend	D3D11_BLEND_ONE
BlendOp	D3D11_BLEND_OP_SUBTRACT
SrcBlendAlpha	D3D11_BLEND_ONE
DestBlendAlpha	D3D11_BLEND_ONE
BlendOpAlpha	c
RenderTargetWriteMask	D3D11_COLOR_WRITE_ENABLE_ALL

The second blend state was used for light accumulation. Use the following parameters for the blend state descriptor:

Blend State Descriptor Parameter	Value
BlendEnable	TRUE
SrcBlend	D3D11_BLEND_ONE
DestBlend	D3D11_BLEND_ONE
BlendOp	D3D11_BLEND_OP_ADD
SrcBlendAlpha	D3D11_BLEND_ONE
DestBlendAlpha	D3D11_BLEND_ONE
BlendOpAlpha	D3D11_BLEND_OP_ADD
RenderTargetWriteMask	D3D11_COLOR_WRITE_ENABLE_ALL

How to do it...

We will be using the quarter resolution depth buffer we prepared for the SSAO calculation and write the results into the light accumulation buffer, so you should place SSLR calculation after the light calculation but before postprocessing.

First, we prepare the occlusion texture. Set the first quarter resolution UAV we allocated in the *Getting ready* section along with the quarter resolution depth buffer from the SSAO calculation. The deceleration for the depth buffer in the compute shader is as follows:

```
Texture2D<float> DepthTex : register( t0 );
RWStructuredBuffer<float> OcclusionRW : register( u0 );
```

Use the following constant declaration:

```
cbuffer OcclusionConstants : register( b0 )
{
uint2 Res : packoffset( c0 );
}
```

The compute shader we will be executing is as follows:

```
[numthreads(1024, 1, 1)]
void Occlussion(uint3 dispatchThreadId : SV_DispatchThreadID)
{
  uint3 CurPixel = uint3(dispatchThreadId.x % Res.x,
  dispatchThreadId.x / Res.x, 0);

  // Skip out of bound pixels
  if(CurPixel.y<Res.y)
  {
    // Get the depth
    float curDepth = DepthTex.Load(FullResPixel);

    // Flag anything closer than the sky as occlusion
    OcclusionRW[dispatchThreadId.x].x = curDepth> 0.99;
  }
}
```

Next, you may want to render the clouds into the occlusion texture. If you don't support rendering clouds, you may skip this step. In order to include the clouds in the occlusion texture, you will have to use the first blend state we created in the *Getting ready* section along with the occlusion texture's RTV. Render the clouds and output their alpha value to the red channel from the pixel shader.

With the occlusion texture ready, we proceed to calculate the ray's intensity. Restore the blend state to the original state (blending disabled), set the second quarter resolution RTV, and set the occlusion textures SRV to slot 0. We will be drawing a full screen quad like we did in previous recipes. Set the primitive topology to triangle strip and call a draw with four vertices and offset 0. Similar to previous recipes, we will use the following constant positions and UV values in the vertex shader to position the quad:

```
static const float2 arrBasePos[4] = {
    float2(-1.0, 1.0),
    float2(1.0, 1.0),
    float2(-1.0, -1.0),
    float2(1.0, -1.0),
};
static const float2 arrUV[4] = {
    float2(0.0, 0.0),
    float2(1.0, 0.0),
    float2(0.0, 1.0),
    float2(1.0, 1.0),
};
```

The vertex shader output structure is defined as follows:

```
struct VS_OUTPUT
{
float4 Position : SV_Position;
float2 UV: TEXCOORD0;
};
```

The following vertex shader will be used to position the vertices:

```
VS_OUTPUT RayTraceVS(uintVertexID : SV_VertexID )
{
    VS_OUTPUT Output;

Output.Position = float4(arrBasePos[VertexID].xy, 1.0, 1.0);
Output.UV = arrUV[VertexID].xy;

return Output;
}
```

During the pixel shader execution, we will be sampling the occlusion texture using a point sampler with the help of the following deceleration:

```
Texture2D<float> OcclusionTex : register( t0 );
SamplerState PointSampler : register( s0 );
```

In addition, we will be using the following constant buffer:

```
cbuffer RayTraceConstants : register( b0 )
{
  float2 SunPos : packoffset( c0 );
  float InitDecay : packoffset( c0.z );
  float DistDecay : packoffset( c0.w );
  float3 RayColor : packoffset( c1 );


}
```

The following pixel shader is used for calculating the light ray's intensity:

```
float4 RayTracetPS( DS_OUTPUT In ) : SV_TARGET
{
  // Initial intensity
  float rayIntensity = OcclusionTex.Sample( PointSampler, UV );

  // Find the ray march delta and offset
  float2 dirToSun = (SunPos - UV);
  float2 rayDelta = dirToSun / 7.0;
  float2 rayOffset = dirToSun;

  // Decay value
  float decay = InitDecay;

  // Ray march towards the Sun
  for(int i = 0; i< 8; i++)
  {
    float fCurIntensity = OcclusionTex.Sample( PointSampler, UV +
rayOffset );

    // Sum the intensity taking decay into account
    rayIntensity += fCurIntensity * decay;

    // Update the decay
    decay += DistDecay;

    // Advance to the next position
    rayOffset += rayDelta;
  }

  return float4(rayIntensity, rayIntensityrayIntensityrayIntensity);
}
```

Now that we calculated the light ray's intensity, we can combine them with the values stored in the light accumulation texture. We will be using the same technique to render another full screen quad, only this time we will be using a different pixel shader. Set the second blend state we declared in the *Getting ready* section. Set the light accumulation texture RTV and the light ray's intensity SRV to slot 0. Our pixel shader will be using a linear sampler to sample the intensity texture using the following deceleration:

```
Texture2D<float> IntensityTex : register( t0 );
Sampler StateLinearSampler : register( s1 );
```

We will be using the same pixel shader constant buffer, so make that sure it is still bound. The pixel shader code is as follows:

```
float4 CombinePS( DS_OUTPUT In ) : SV_TARGET
{
  // Ge the ray intensity
  float rayIntensity = OcclusionTex.Sample( LinearSampler, UV );

  // Return the color scaled by the intensity
  return float4(RayColor * rayIntensity, 1.0);
}
```

How it works...

In both, forward and deferred lighting, pixels were only getting lit by light rays reflected from surfaces to the camera. It is important to understand that the Blinn-Phong lighting model assumes that all light travels in a vacuum, which is obviously not the case in real life. In reality, air contains different particles that interact with the light rays in two ways. Those particles can either absorb the light, which means that the ray doesn't travel anywhere after it hits the particle, or they can reflect or scatter the light in a different direction than the one it was traveling in before it hit the particle.

Sun rays are the result of light rays changing direction toward the camera due to scattering. Since we are not going to get this effect using the Blinn-Phong lighting equations introduced in *Chapter 1, Forward Lighting*, and *Chapter 2, Deferred Shading*, we need to calculate the scattering effect separately and add it on top of the other accumulated light values. SSLR lets us calculate the scattering contribution in three steps: occlusion, tracing the rays, and combining the rays with the light accumulation texture.

There are two main contributors to the occlusion map. The first is the scene's opaque geometry and the second is the clouds. Any pixel not covered by either of the two can potentially be a source for the light rays. We want pixels that are fully occluded to be set to black while pixels that are fully unoccluded be set to white. Pixels that are partially occluded due to clouds should have an intermediate value. Occlusions from the opaque geometry are generated based on the SSAO quarter scale depth values using a compute shader.

If a pixel has a depth lower than one, which is the value for the back plane, it means that some opaque geometry was written into it and it is occluded. Once the opaque geometry is in, we need to add the cloud's occlusion on top to get the final occlusion values. If you do not support cloud rendering, then you can skip this step. The clouds you render into the occlusion map have to match the ones you rendered into the light accumulation texture, otherwise the Sun rays may end up going through the clouds, which will look wrong. Since the clouds' occlusion is supposed to reduce the value in the occlusion texture, clouds output a higher value of how thicker they are and we use subtraction operation for the blending. This way the clouds reduce the value in the occlusion texture, which increases the occlusion for pixels they affect.

Once the first step is completed and the occlusion texture is ready, we move on to the next step where we find the per-pixel light ray's intensity values. We extrude the light rays using ray marching, which is similar to the way we used ray tracing for reflections, only this time we don't want to do any sort of intersection testing. For each pixel, we ray march in 2D from the pixel position towards the Sun position and accumulate the scaled nonoccluded values we find along the way. Pixels that come across more nonoccluded pixels along the way will receive a larger contribution and the light ray will show a brighter picture.

At this point, we are ready for the final stage where we combine the light rays with the values in the light accumulation buffer. We render a full screen quad that covers the entire screen. In the pixel shader, values from the light ray's intensity values found in the previous stage are sampled using a linear filter. To get the final pixel value, we multiply the sampled intensity by the scaled rays color and write the result into the light accumulation buffer using the same additive blending we used for deferred lighting.

6
Environment Effects

In this chapter we will cover:

- ▶ Dynamic decals
- ▶ Distance/Height-based fog
- ▶ Rain

Introduction

Lighting and postprocessing are two of the most important fields in 3D rendering. A scene that is properly lit with the right postprocessing effects can produce beautiful results, but not every element you may want to render can be represented by a mesh you can load from file. One common scenario where a mesh loaded from a file cannot be used is when the mesh has to be dynamically generated at runtime based on some unpredictable event, like a mesh that represents damage to the scene. Another good example is when rendering an element which is too complex to be represented by triangles, like smoke and fire.

In this chapter, we will focus on techniques which implement those various scene elements, which we can't represent with a loaded mesh. The recipes presented in this chapter can be split into three main categories:

- ▶ Dynamic mesh generation: It is used in cases where the mesh shape can't be predicted beforehand, but once the mesh is prepared it no longer changes
- ▶ Depth-based rendering: It is used when we want to represent a medium that affects light as it travels in it based on the distance the light had to cover in the medium on its way to the camera
- ▶ Particle system: It is used when the shape we want to render is too complicated or too dynamic to represent with a mesh made of triangles

It's important to remember that even though each technique presented in this chapter may seem like a solution to a very specific problem, the technique used in each recipe can be modified to render a large range of effects and dynamic elements.

Dynamic decals

Dynamic decals are probably the most common way to indicate some interaction with the environment. Decals can represent anything from a bullet impact location to vehicle tire skid marks. A common way to render decals is to alpha blend them on top of the environment, so the pixels covered by the decal can be changed to the decal values while keeping the rest of the scene the same. Because the decals are applied on top of the scene, it is important for each decal to fit perfectly on top of the scene without causing any artifacts. There are two common ways to render decals: projection of the decal using the scenes depth values or by rendering a mesh representing the decal. Both of these methods have pros and cons, but the second one offers an opportunity to use the ability to stream out vertices, which makes this option more suitable for this book.

As an example, the following image shows the same scene first without any decals (left side of the image) and then which decals applied to various positions (right side of the image):

This recipe will focus on decals applied in response to an intersection of a ray with a mesh, which is a very common scenario resulting in decal generation. In order to keep this recipe focused on the decal generation, the ray intersection process is not going to be covered.

Getting ready

We are going to prepare a separate set of assets for the decal generation process and for the actual rendering. First, we will cover the assets required for generating the decals. Create the vertex shader constant buffer using the following buffer descriptor:

Constant buffer descriptor parameter	Value
Usage	D3D11_USAGE_DYNAMIC
BindFlags	D3D11_BIND_CONSTANT_BUFFER
CPUAccessFlags	D3D11_CPU_ACCESS_WRITE
ByteWidth	64

The rest of the descriptor fields should be set to zero.

Create the geometry shader constant buffer using the following buffer descriptor:

Constant buffer descriptor parameter	Value
Usage	D3D11_USAGE_DYNAMIC
BindFlags	D3D11_BIND_CONSTANT_BUFFER
CPUAccessFlags	D3D11_CPU_ACCESS_WRITE
ByteWidth	128

The reset of the descriptor fields should be set to zero.

Our geometry shader will be using the stream out feature introduced by DirectX 10. When a geometry shader outputs vertices into a stream, it has to be created with additional information that describes its output. Create the geometry shader with the function:

```
HRESULT CreateGeometryShaderWithStreamOutput(
    [in]    const void *pShaderBytecode,
    [in]    SIZE_T BytecodeLength,
    [in]    const D3D11_SO_DECLARATION_ENTRY *pSODeclaration,
    [in]    UINT NumEntries,
    [in]    const UINT *pBufferStrides,
    [in]    UINT NumStrides,
    [in]    UINT RasterizedStream,
    [in]    ID3D11ClassLinkage *pClassLinkage,
    [out]   ID3D11GeometryShader **ppGeometryShader
);
```

The third parameter to create the function is an array which we declare as follows:

```
D3D11_SO_DECLARATION_ENTRY pStreamOutDecl[] =
{
    { 0, "POSITION", 0, 0, 3, 0 },
    { 0, "NORMAL", 0, 0, 3, 0 },
    { 0, "TEXCOORD", 0, 0, 2, 0 },
};
```

This array contains the output vertices structure where each entry contains: stream number, semantic name, semantic index, start component, component count, and output slot. The number of entries is going to be 3.

Use a stride of 8 for the fifth parameter (3 position floats, 3 normal floats, and 2 texture coordinate floats).

Next, we will create a vertex buffer that will store the decal vertices. As we can't support unlimited amount of decals, the size of the vertex buffer will dictate the maximum amount of decals in any given time. For this recipe, we will support up to 1024 decal vertices. Create the vertex buffer using the following descriptor:

Vertex buffer descriptor parameter	Value
Usage	D3D11_USAGE_DEFAULT
BindFlags	D3D11_BIND_VERTEX_BUFFER | D3D11_BIND_STREAM_OUTPUT
ByteWidth	1024 * sizeof(<Vertex Structure>) * sizeof(float)

The reset of the descriptor fields should be set to zero.

In addition to all the preceding values, we will be using the depth stencil state with depth and stencil disabled we declared for postprocessing.

Another new asset we will need is a stream out statistics query to keep track of the geometry shader output size. Use the following descriptor when you call `CreateQuery`:

Query descriptor parameter	Value
Query	D3D11_QUERY_SO_STATISTICS

The reset of the descriptor fields should be set to zero.

Rendering the decals can be handled in a very similar way to how we render the scene into the GBuffer, so we won't need new shaders or constant buffers. We will however need two new states. Create a rasterizer state using the following descriptor:

Rasterizer state descriptor parameter	Value
FillMode	D3D11_FILL_SOLID
CullMode	D3D11_CULL_BACK
FrontCounterClockwise	FALSE
DepthBias	This should be set to a negative integer value-
DepthBiasClamp	D3D11_DEFAULT_DEPTH_BIAS_CLAMP

Rasterizer state descriptor parameter	Value
SlopeScaledDepthBias	This should be set to a negative float value
DepthClipEnable	FALSE
ScissorEnable	FALSE
MultisampleEnable	FALSE
AntialiasedLineEnable	FALSE

Create a blend state with the following descriptor:

Blend state descriptor parameter	Value
BlendEnable	TRUE
SrcBlend	D3D11_BLEND_SRC_ALPHA
DestBlend	D3D11_BLEND_INV_SRC_ALPHA
BlendOp	D3D11_BLEND_OP_ADD
SrcBlendAlpha	D3D11_BLEND_SRC_ALPHA
DestBlendAlpha	D3D11_BLEND_INV_SRC_ALPHA
BlendOpAlpha	D3D11_BLEND_OP_ADD
RenderTargetWriteMask	D3D11_COLOR_WRITE_ENABLE_ALL

Finally, we will render the decals with a texture that represents puncture damage to a concrete surface. In the image shown at introduction of this recipe, the following texture was used:

How to do it...

For the decal generation, this recipe assumes that there is some mechanism in place which handles a ray intersection with the scene to find the mesh the decal is going to be applied to. In addition, we will need the intersection of world position and normal. We will be generating the decals by drawing the mesh that was intersected by the ray, so make sure the next steps are handled inside the frame's context begin and end calls.

The first step in generating the corresponding decal is to prepare the constant buffers. The vertex shader constant buffer should be set to the same world matrix used by the mesh the decal is going to affect. This buffer is declared in the shader as follows:

```
cbuffer DecalGenVSConstants: register( b0 )
{
    float4x4 World : packoffset( c0 );
}
```

For the geometry shader constant buffer, we will use the following structure:

```
cbuffer DecalGenGSConstants: register( b0 )
{
    float4 ArrClipPlanes[6]    : packoffset( c0 );
    float2 DecalSize        : packoffset( c6 );
    float3 HitNorm        : packoffset( c7 );
}
```

Decal size is the size of the decal in world units. Clip planes represent the 3D volume affected by the decal. You can generate those six clip planes by creating an orthogonal system from the mesh intersection normal (for a full explanation, check the *Cascaded Shadows* recipe). The six planes are split into pairs where each pair uses one of the three axes from the orthogonal system. One plane out of each pair uses the axis vector as its normal while the other plane uses the negated axis vector as its normal. To get the planes distance (the W component), subtract half the decal size from the dot product between the plane normal and the intersection position.

Once the constant buffer is ready, we will draw the mesh that got hit by the ray. Bind the stream out buffer using the `SOSetTargets` function. Make sure you set the offset value to `(UINT) (-1)` when calling `SOSetTargets`, so the new decal vertices get appended after the last data added to the buffer. Bind the depth stencil state with depth and stencil disabled (this will prevent D3D from complaining about the depth state not being compatible with the stream out operation). Finally, call begin for statistics query and draw the mesh.

Our vertex and geometry shaders will use the following vertex shader deceleration for both input and output:

```
struct VERT_INPUT_OUTPUT
{
    float3 Pos    : POSITION;
    float3 Norm    : NORMAL;
    float2 Tex    : TEXCOORD0;
};
```

We will be using the following vertex shader which transforms the positions and normal into world space:

```
VERT_INPUT_OUTPUT DecalGenVS(VERT_INPUT_OUTPUT input)
{
    VERT_INPUT_OUTPUT output;

    output.Pos = mul(float4(input.Pos, 1.0), World).xyz;
    output.Norm = mul(input.Norm, (float3x3)World);
    output.Tex = input.Tex;

    return output;
}
```

Once the vertices are in world space, our geometry shader can clip each one of the mesh triangles with the six clip planes which will result in the decal triangles. Clipping those triangles will be handled by two helper functions. The following code finds the intersection between a single plane and a segment:

```
void PlaneSegIntersec(float4 p1, float3 norm1, float4 p2,
float3 norm2, float4 plane, out float4 intersectPos, out float3
intersectNorm)
{
    float3 segDir = p2.xyz - p1.xyz;
    float segDist = length(segDir);
    segDir = segDir / segDist;
    float fDenom = dot(plane.xyz, segDir);

    float fDist = -dot(plane, p1) / fDenom;
    intersectPos = float4(p1.xyz + fDist * segDir, 1.0);

    // Calculate the normal
    intersectNorm = lerp(norm1, norm2, fDist / segDist);
    intersectNorm = normalize(intersectNorm);
}
```

The second helper function will clip a single triangle with a single plane using the following code:

```
void PolyPlane(float4 arrVerts[MAX_NEW_VERT], float3 arrNormals[MAX_
NEW_VERT], float arrDot[MAX_NEW_VERT], uint iNumVerts, float4 plane,
out float4 arrNewVerts[MAX_NEW_VERT], out float3 arrNewNormals[MAX_
NEW_VERT], out uint iCurNewVert)
{
    arrNewVerts = (float4[MAX_NEW_VERT])100000.0f;
    iCurNewVert = 0;
```

```
    for(uint i=0; i < iNumVerts; i++)
    {
        if(arrDot[i] >= 0)
        {
            arrNewVerts[iCurNewVert] = arrVerts[i];
            arrNewNormals[iCurNewVert] = arrNormals[i];
            iCurNewVert++;

            if(arrDot[(i+1)%iNumVerts] < 0)
            {
                PlaneSegIntersec(arrVerts[i], arrNormals[i],
                arrVerts[(i+1)%iNumVerts], arrNormals[(i+1)%iNumVerts],
                plane, arrNewVerts[iCurNewVert],
                arrNewNormals[iCurNewVert]);
                iCurNewVert++;
            }
        }
        else if(arrDot[(i+1)%iNumVerts] >= 0)
        {
            PlaneSegIntersec(arrVerts[i], arrNormals[i],
            arrVerts[(i+1)%iNumVerts], arrNormals[(i+1)%iNumVerts],
            plane, arrNewVerts[iCurNewVert], arrNewNormals[iCurNewVert]);
            iCurNewVert++;
        }
    }
}
```

Finally, the code for the geometry shader entry point is as follows:

```
[maxvertexcount(12)] // Max 4 triangles x 3 vertices
void DecalGenGS(triangle VERT_INPUT_OUTPUT input[3], inout
TriangleStream<VERT_INPUT_OUTPUT> TriStream)
{
    uint nNumVerts = 0;
    float4 arrNewVerts[MAX_NEW_VERT] = (float4[MAX_NEW_VERT])100000.0;
    float3 arrNewNormals[MAX_NEW_VERT] = (float3[MAX_NEW_VERT])0.0;
    uint iIn[MAX_NEW_VERT] = (uint[MAX_NEW_VERT])0;
    float arrDot[MAX_NEW_VERT] = (float[MAX_NEW_VERT])0;

    arrNewVerts[0] = float4(input[0].Pos, 1.0);
    arrNewVerts[1] = float4(input[1].Pos, 1.0);
    arrNewVerts[2] = float4(input[2].Pos, 1.0);
    arrNewNormals[0] = input[0].Norm;
    arrNewNormals[1] = input[1].Norm;
    arrNewNormals[2] = input[2].Norm;
```

```
// Make sure the triangle is not facing away from the hit ray
float3 AB = arrNewVerts[1].xyz - arrNewVerts[0].xyz;
float3 AC = arrNewVerts[2].xyz - arrNewVerts[0].xyz;
float3 faceNorm = cross(AB, AC);
float fDot = dot(faceNorm, HitNorm);
nNumVerts = 3 * (fDot > 0.01);

// Clip the triangle with each one of the planes
for(uint iCurPlane=0; iCurPlane < 6; iCurPlane++)
{
   // First check the cull status for each vertex
   for(uint i=0; i < MAX_NEW_VERT; i++ )
   {
      arrDot[i] = dot(ArrClipPlanes[iCurPlane], arrNewVerts[i]);
   }

   // Calculate the new vertices based on the culling status
   uint nNewNumVerts = 0;
   PolyPlane(arrNewVerts, arrNewNormals, arrDot, nNumVerts,
   ArrClipPlanes[iCurPlane], arrNewVerts, arrNewNormals,
   nNewNumVerts);
   nNumVerts = nNewNumVerts;
}

VERT_INPUT_OUTPUT output = (VERT_INPUT_OUTPUT)0;

// Add the new triangles to the stream
for(uint nCurVert = 1; nCurVert < nNumVerts-1 && nNumVerts > 0;
nCurVert++)
{
   output.Pos = arrNewVerts[0].xyz;
   output.Norm = arrNewNormals[0];
   output.Tex.x = dot(arrNewVerts[0], ArrClipPlanes[1]);
   output.Tex.y = dot(arrNewVerts[0], ArrClipPlanes[3]);
   output.Tex = output.Tex / DecalSize;
   TriStream.Append( output );
   output.Pos = arrNewVerts[nCurVert].xyz;
   output.Norm = arrNewNormals[nCurVert];
   output.Tex.x = dot(arrNewVerts[nCurVert], ArrClipPlanes[1]);
   output.Tex.y = dot(arrNewVerts[nCurVert], ArrClipPlanes[3]);
   output.Tex = output.Tex / DecalSize;
   TriStream.Append( output );
   output.Pos = arrNewVerts[nCurVert+1].xyz;
   output.Norm = arrNewNormals[nCurVert+1];
```

```
            output.Tex.x = dot(arrNewVerts[nCurVert+1], ArrClipPlanes[1]);
            output.Tex.y = dot(arrNewVerts[nCurVert+1], ArrClipPlanes[3]);
            output.Tex = output.Tex / DecalSize;
            TriStream.Append( output );
            TriStream.RestartStrip();
        }
    }
```

Make sure to end the query immediately after the draw call.

At this point, our stream out buffer contains the decal mesh information if there was enough room left. Before any drawing can take place, we need to wait for the query results to be ready so we can read them back from the GPU to the CPU. Use `GetData` with a `D3D11_QUERY_DATA_SO_STATISTICS` structure in every frame until the results are ready and the structure is filled. If the values in the structure of `NumPrimitivesWritten` and `PrimitivesStorageNeeded` do not match, it means that the buffer didn't have enough room for the new decal. If the buffer is out of room, you have to repeat the process and add the decal to the beginning of the buffer overwriting older decals (use the value of 0 when setting the buffer with `SOSetTargets` to overwrite). Once the decal was successfully added, you need to store the amount of vertices that where added and their position in the buffer in order to render only the active decals.

Rendering the decals is handled separately from their generation, as we would like to render them in every frame, but only generate new decals as needed. We are going to render the decals into the GBuffer so they get lit together with the mesh they got generated from. Set the blend and rasterizer states we prepared in the *Getting ready...* section so that the decals can alpha blend and avoid Z fighting issues.

Depending on the content of the decals vertex buffer, you may need to render the decals with either one or two draw calls. Each time a decal is added to the vertex buffer, we have to store the value returned by the stream out query. Keeping track over these numbers lets us track the beginning and end range of the decals at every given moment. The following examples show the three scenarios we have to deal with when rendering the decals:

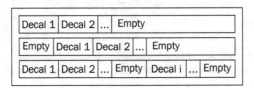

The top vertex buffer contains decals from the start of the buffer up to some point in the middle of the buffer. In the middle vertex buffer, we see a gap at the start of the vertex buffer. This happens right when an overflow is detected and some room is freed at the start of the buffer before the last decal is re-added. Finally, the bottom vertex buffer contains some decals at the start of the buffer, then a gap where new decals are added and more decals after that gap. This scenario happens as soon as the first decal is added after an overflow is detected and additional overwrite older decals in the middle of the buffer. Both top and middle scenarios are rendered with a single draw call while the bottom scenario requires two separate draw calls (one for each range in the buffer).

As the decals get rendered in the same way we render the rest of the scene, you can use the same vertex shader and constant buffer as you would use for any other mesh. The only difference would be in the pixel shader where you will need to copy the alpha value sampled from the decal texture into output color 0.

How it works...

Decals are handled in two different steps. First, the decals mesh is generated from the ray intersection position, the mesh that got intersected and the decals influence volume (defined by the six planes). The mesh is then stored as a vertex buffer for rendering. Once the decals vertex buffer is ready, that buffer is used to render each one of the decals on top of the mesh the decal was generated from.

Up until DirectX 10 was released, decals were generated entirely on the CPU. The main disadvantage of that approach was performance. Performance wise, the original approach performance depended on the amount of triangles in the mesh the decal was applied to, as each triangle had to be processed for intersection with the decal bounds. Another problem with the old approach was getting CPU access to the mesh memory. As the mesh is rendered by the GPU, the CPU doesn't have a direct access to the mesh data getting rendered. Unless a copy of each mesh was stored on the CPUs main memory, the mesh index and vertex buffers had to be locked before the CPU could access them. This lock process takes time which made the whole process slower.

Since the introduction of Geometry Shaders and Stream Out operations as part of DirectX 10 feature set, a new possibility for dynamic decals generation emerged. Instead of handling the hit mesh clipping calculation on the GPU, DirectX 10 made it possible to move this calculation to the Geometry shader and output the result directly into a vertex buffer. That vertex buffer can then be used for rendering the decals. With this new approach, most of the work is handled by the GPU and all the CPU has to do is find the ray intersection position and normal (this work is usually handled by the physics system).

Our geometry shader handles the triangle clipping by clipping each one of the mesh triangles against all six planes. During this operation, we handle the triangle as a convex polygon made out of a set of points. The geometry shaders entry point starts off by copying the triangles' three positions into an array that will represent our convex polygon. During the first for loop in the entry point function, each plane clips the convex outline of the convex polygon against one plane and updates the polygon points based on the results. As each iteration clips the result from the previous iteration, the input triangle can end up as a convex polygon with up to six points, or get completely clipped out. The following images show a possible clipping scenario (represented in 2D for simplification) where the leftmost image features the input triangle and the rightmost image features the resulting convex polygon which is made of six points:

Clipping the polygon is handled by the two helper functions: `PolyPlane` and `PlaneSegIntersec`. Inside `PolyPlane`, we iterate over the current set of polygon points treating neighboring points as the start and end points of one segment of the polygon outline. `PlaneSegIntersec` is then used to find the intersection position on the segment. If both points are inside the plane's positive side, the first out of the two points is selected. Once the polygon is clipped against all planes, the resulting polygon is going to be part of the decal mesh.

As you can see in the example scenario, the resulting polygon is too complex to render and needs to be split into triangles. This is where the second loop in the entry point comes into play. The way we triangulate the polygon is to pick the first point as the base for all triangles. We then add two neighboring points to the base point and generate a set of triangles that can be added to the vertex buffer. The following image shows how the polygon resulting from the first loop is triangulated in the second loop:

Once each triangle from the input mesh is clipped, triangulated, and stored into the vertex buffer, the vertex buffer is ready for rendering, but we can't draw it until we know how many triangles got added on the CPU side. DirectX 10 added a new draw function that tries to help with this issue which is called `DrawAuto` that lets us draw the result of the stream out operation without knowing how many vertices got added. There is one case that requires special attention: when we are overwriting decals in the middle of the buffer, `DrawAuto` will only draw triangles from the start of the buffer up to the last triangle added. In this case an additional call to `Draw` is required to render the second portion of the buffer towards the end.

Before we can conclude this recipe, here is a short explanation to the depth bias we use while rendering the decals. Each triangle in the decal mesh should fully overlap with at least part of a triangle from the original mesh. As we render the decals after the scene, the depth value for each decal pixel should be equal (within precision limitations) to the value stored in the depth buffer when we rendered the parent mesh. The depth bias we use pushes the decal depth toward the camera, so it would pass the depth test when its parent mesh is visible and fail when the parent is hidden by a mesh closer to the camera.

There's more...

Outputting the decals' positions and normal values in world space is very convenient, but will not work properly when the decals are applied to a mesh that moves. When rendering a dynamic mesh that moves over time, its vertices are stored in a local space and are transformed to their correct world space by a different world matrix every time they move. Adapting this recipe to handle dynamic meshes is pretty straightforward.

When generating the decal, don't transform the positions and normal to world space in the vertex shader. Instead, inverse the meshes world matrix and use it to transform the coordinate system used for generating the clip planes into local space. This way the decal triangles will be clipped and stored into the vertex buffer in local space.

When rendering the decals, you will need a separate `Draw` call for every decal that belongs to a dynamic mesh. Just set the same world matrix the decal parent mesh was using before rendering the decal.

Another common solution is to allocate a separate vertex buffer that will only contain the vertices of that dynamic mesh. This way the vertices are still stored in local space, but can be rendered in the same way the static mesh vertex buffer was handled.

Distance/Height-based fog

Distance/Height-based fog is an approximation to the fog you would normally see outdoors. Even in the clearest of days, you should be able to see some fog far in the distance. The main benefit of adding the fog effect is that it helps the viewer estimate how far different elements in the scene are based on the amount of fog covering them. In addition to the realism this effect adds, it has the additional benefit of hiding the end of the visible range. Without fog to cover the far plane, it becomes easier to notice when far scene elements are clipped by the cameras far plane.

By tuning the height of the fog you can also add a darker atmosphere to your scene as demonstrated by the following image:

This recipe will demonstrate how distance/height-based fog can be added to our deferred directional light calculation. See the *How it works...* section for details about adding the effect to other elements of your rendering code.

Getting ready

We will be passing additional fog specific parameters to the directional light's pixel shader through a new constant buffer. The reason for separating the fog values into their own constant buffer is to allow the same parameters to be used by any other shader that takes fog into account. To create the new constant buffer use the following buffer descriptor:

Constant buffer descriptor parameter	Value
Usage	D3D11_USAGE_DYNAMIC
BindFlags	D3D11_BIND_CONSTANT_BUFFER
CPUAccessFlags	D3D11_CPU_ACCESS_WRITE
ByteWidth	48

The reset of the descriptor fields should be set to zero.

All the fog calculations will be handled in the deferred directional light pixel shader.

How to do it...

Our new fog constant buffer is declared in the pixel shader as follows:

```
cbuffer cbFog : register( b2 )
{
    float3 FogColor            : packoffset( c0 );
    float FogStartDepth        : packoffset( c0.w );
    float3 FogHighlightColor   : packoffset( c1 );
    float FogGlobalDensity     : packoffset( c1.w );
    float3 FogSunDir           : packoffset( c2 );
    FogHeightFalloff           : packoffset( c2.w );
}
```

The helper function used for calculating the fog is as follows:

```
float3 ApplyFog(float3 originalColor, float eyePosY, float3
eyeToPixel)
{
 float pixelDist = length( eyeToPixel );
 float3 eyeToPixelNorm = eyeToPixel / pixelDist;

 // Find the fog staring distance to pixel distance
 float fogDist = max(pixelDist - FogStartDist, 0.0);

 // Distance based fog intensity
 float fogHeightDensityAtViewer = exp( -FogHeightFalloff * eyePosY );
 float fogDistInt = fogDist * fogHeightDensityAtViewer;

 // Height based fog intensity
 float eyeToPixelY = eyeToPixel.y * ( fogDist / pixelDist );
 float t = FogHeightFalloff * eyeToPixelY;
 const float thresholdT = 0.01;
 float fogHeightInt = abs( t ) > thresholdT ?
  ( 1.0 - exp( -t ) ) / t : 1.0;

 // Combine both factors to get the final factor
 float fogFinalFactor = exp( -FogGlobalDensity * fogDistInt *
fogHeightInt );

 // Find the sun highlight and use it to blend the fog color
 float sunHighlightFactor = saturate(dot(eyeToPixelNorm, FogSunDir));
 sunHighlightFactor = pow(sunHighlightFactor, 8.0);
```

```
    float3 fogFinalColor = lerp(FogColor, FogHighlightColor,
sunHighlightFactor);

    return lerp(fogFinalColor, originalColor, fogFinalFactor);
}
```

The `Applyfog` function takes the color without fog along with the camera height and the vector from the camera to the pixel the color belongs to and returns the pixel color with fog. To add fog to the deferred directional light, change the directional entry point to the following code:

```
float4 DirLightPS(VS_OUTPUT In) : SV_TARGET
{
    // Unpack the GBuffer
    float2 uv = In.Position.xy;//In.UV.xy;
    SURFACE_DATA gbd = UnpackGBuffer_Loc(int3(uv, 0));

    // Convert the data into the material structure
    Material mat;
    MaterialFromGBuffer(gbd, mat);

    // Reconstruct the world position
    float2 cpPos = In.UV.xy * float2(2.0, -2.0) - float2(1.0, -1.0);
    float3 position = CalcWorldPos(cpPos, gbd.LinearDepth);

    // Get the AO value
    float ao = AOTexture.Sample(LinearSampler, In.UV);

    // Calculate the light contribution
    float4 finalColor;
    finalColor.xyz = CalcAmbient(mat.normal, mat.diffuseColor.xyz) * ao;
    finalColor.xyz += CalcDirectional(position, mat);
    finalColor.w = 1.0;

    // Apply the fog to the final color
    float3 eyeToPixel = position - EyePosition;
    finalColor.xyz = ApplyFog(finalColor.xyz, EyePosition.y,
eyeToPixel);

    return finalColor;
}
```

With this change, we apply the fog on top of the lit pixels color and return it to the light accumulation buffer.

How it works...

Fog is probably the first volumetric effect implemented using a programmable pixel shader as those became commonly supported by GPUs. Originally, fog was implemented in hardware (fixed pipeline) and only took distance into account. As GPUs became more powerful, the hardware distance based fog was replaced by a programmable version that also took into account things such as height and sun effect.

In reality, fog is just particles in the air that absorb and reflect light. A ray of light traveling from a position in the scene travels, the camera interacts with the fog particles, and gets changed based on those interactions. The further this ray has to travel before it reaches the camera, the larger the chance is that this ray will get either partially or fully absorbed. In addition to absorption, a ray traveling in a different direction may get reflected towards the camera and add to the intensity of the original ray. Based on the amount of particles in the air and the distance a ray has to travel, the light reaching our camera may contain more reflection and less of the original ray which leads to a homogenous color we perceive as fog.

The parameters used in the fog calculation are:

- `FogColor`: The fog base color (this color's brightness should match the overall intensity so it won't get blown by the bloom)

- `FogStartDistance`: The distance from the camera at which the fog starts to blend in

- `FogHighlightColor`: The color used for highlighting pixels with pixel to camera vector that is close to parallel with the camera to sun vector

- `FogGlobalDensity`: Density factor for the fog (the higher this is the denser the fog will be)

- `FogSunDir`: Normalized sun direction

- `FogHeightFalloff`: Height falloff value (the higher this value, the lower is the height at which the fog disappears will be)

When tuning the fog values, make sure the ambient colors match the fog. This type of fog is designed for outdoor environments, so you should probably disable it when lighting interiors.

 You may have noticed that the fog requires the sun direction. We already store the inversed sun direction for the directional light calculation. You can remove that value from the directional light constant buffer and use the fog vector instead to avoid the duplicate values

This recipe implements the fog using the exponential function. The reason for using the exponent function is because of its asymptote on the negative side of its graph. Our fog implementation uses that asymptote to blend the fog in from the starting distances. As a reminder, the exponent function graph is as follows:

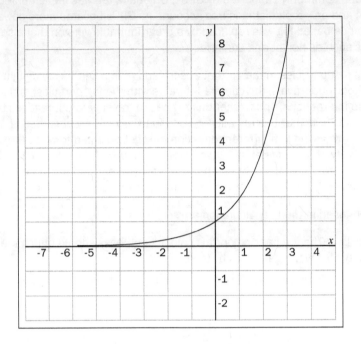

The `ApplyFog` function starts off by finding the distance our ray traveled in the fog (`fogDepth`). In order to take the fog's height into account, we also look for the lowest height between the camera and the pixel we apply the fog to which we then use to find how far our ray travels vertically inside the fog (`fogHeight`). Both distance values are negated and multiplied by the fog density to be used as the exponent. The reason we negate the distance values is because it's more convenient to use the negative side of the exponential functions graph which is limited to the range 0 to 1. As the function equals 1 when the exponent is 0, we have to invert the results (stored in `fogFactors`).

At this point we have one factor for the height which gets larger the further the ray travels vertically into the fog and a factor that gets larger the further the ray travels in the fog in any direction. By multiplying both factors with each other we get the combined fog effect on the ray: the higher the result is, the more the original ray got absorbed and light got reflected towards the camera in its direction (this is stored in `fogFinalFactor`).

Before we can compute the final color value, we need to find the fog's color based on the camera and sun direction. We assume that the sun intensity is high enough to get more of its light rays reflected towards the camera direction and sun direction are close to parallel. We use the dot product between the two to determine the angle and narrow the result by raising it to the power of 8 (the result is stored in `sunHighlightFactor`). The result is used to lerp between the fog base color and the fog color highlighted by the sun.

Finally, we use the fog factor to linearly interpolate between the input color and the fog color. The resulting color is then returned from the helper function and stored into the light accumulation buffer.

As you can see, the changes to the directional light entry point are very minor as most of the work is handled inside the helper function `ApplyFog`. Adding the fog calculation to the rest of the deferred and forward light sources should be pretty straightforward. One thing to take into consideration is that fog also has to be applied to scene elements that don't get lit, like the sky or emissive elements. Again, all you have to do is call `ApplyFog` to get the final color with the fog effect.

Rain

Particle systems are a very common technique used when rendering something with a shape that's either too complex to render using single mesh or when there are many simple shapes that move in a complex pattern. Fire is a good example for the first case, because the flames produce a random shape that can't be modeled beforehand. On the other hand, rainfall is made out of many simple streaks which move in a complex pattern, which is an example for the second case.

In this recipe, we will concentrate on rain because it requires a relatively simple simulation which is mostly effected by gravity. With a slight modification, the same simulation can be extended to support snow and steroid showers. Other types of particle effects such as smoke and fire, can also be handled entirely on the GPU, but the simulation required is going to be much more complicated.

The following screenshot features a scene with rain:

Getting ready

Similar to the way we handled the highlights for the bokeh effect, we will be using a buffer that will be updated through a compute shader and rendered as a vertex buffer. Allocate the buffer with the following descriptor values:

Buffer descriptor parameter	Value
BindFlags	D3D11_BIND_UNORDERED_ACCESS | D3D11_BIND_SHADER_RESOURCE
StructureByteStride	7 * (size of float)
ByteWidth	128 * 128 * StructureByteStride
MiscFlags	D3D11_RESOURCE_MISC_BUFFER_STRUCTURED

Before creating this buffer, make sure to initialize all the floats to a very low negative value to force the simulation to initialize on the first update.

During the simulation, we will be using a depth texture which is allocated with the following descriptor values:

Texture descriptor parameter	Value
Width	128
Height	128
MipLevels	1
ArraySize	1
Format	DXGI_FORMAT_R32_TYPELESS
BindFlags	D3D11_BIND_DEPTH_STENCIL | D3D11_BIND_SHADER_RESOURCE

The reset of the descriptor fields should be set to zero.

We will be rendering into the texture with the following depth stencil view descriptor:

DSV descriptor parameter	Value
Format	DXGI_FORMAT_D32_FLOAT
ViewDimension	D3D11_DSV_DIMENSION_TEXTURE2D

The reset of the descriptor fields should be set to zero.

During the simulation, we will be sampling from the depth texture with the following shader resource view descriptor:

SRV descriptor parameter	Value
Format	DXGI_FORMAT_R32_FLOAT
ViewDimension	D3D11_SRV_DIMENSION_TEXTURE2D
Texture2D.MipLevels	1

The reset of the descriptor fields should be set to zero.

We will need a constant buffer that will be used while rendering into the depth texture with the following descriptor parameters:

Constant buffer descriptor parameter	Value
Usage	D3D11_USAGE_DYNAMIC
BindFlags	D3D11_BIND_CONSTANT_BUFFER
CPUAccessFlags	D3D11_CPU_ACCESS_WRITE
ByteWidth	64

The reset of the descriptor fields should be set to zero.

During the simulation we will be using another constant buffer with the following descriptor parameters:

Constant buffer descriptor parameter	Value
Usage	D3D11_USAGE_DYNAMIC
BindFlags	D3D11_BIND_CONSTANT_BUFFER
CPUAccessFlags	D3D11_CPU_ACCESS_WRITE
ByteWidth	112

The reset of the descriptor fields should be set to zero.

During the rain simulation, we will be sampling from a noise texture to get random values. This texture size should be 128x128 and would contain high frequency noise in all four channels. There is no need to generate mipmaps for this texture as we will be sampling only from the full resolution. The following image is an example to how one channel of the noise texture may look like:

In order to use the texture, load it into an SRV.

Rendering the rain will require an additional constant buffer with the following descriptor:

Constant buffer descriptor parameter	Value
Usage	D3D11_USAGE_DYNAMIC
BindFlags	D3D11_BIND_CONSTANT_BUFFER
CPUAccessFlags	D3D11_CPU_ACCESS_WRITE
ByteWidth	96

The reset of the descriptor fields should be set to zero.

As rain is transparent, we will be using an alpha blending state with similar settings to the one we used for the dynamic decals. Create the blend state with the following alpha blend state descriptor values:

Blend state descriptor parameter	Value
BlendEnable	TRUE
SrcBlend	D3D11_BLEND_SRC_ALPHA
DestBlend	D3D11_BLEND_INV_SRC_ALPHA
BlendOp	D3D11_BLEND_OP_ADD
SrcBlendAlpha	D3D11_BLEND_SRC_ALPHA
DestBlendAlpha	D3D11_BLEND_INV_SRC_ALPHA
BlendOpAlpha	D3D11_BLEND_OP_ADD
RenderTargetWriteMask	D3D11_COLOR_WRITE_ENABLE_ALL

Finally, we will be using a 4x128 texture with a full mipmap chain which will give the rain streak an uneven look. The following image is the magnified rain texture:

As with the noise texture, you will have to load the rain streak texture into an SRV.

How to do it...

Before we can simulate the rain particles, we have to update the depth texture with the scene's height values. We will be using an orthographic projection to render the scene depths in a similar way to the way we prepared the shadow maps. Set the DSV and clear the depth to 1. Our vertex shader will use the following constant buffer:

```
cbuffer RainConstants : register( b0 )
{
    float4x4 ToHeight : packoffset( c0 );
}
```

Please read the *How it works...* section for full details on how to prepare the matrix.

We will use the following vertex shader to output the depths:

```
float4 HeightMapVS(float4 Pos : POSITION) : SV_POSITION
{
    return mul(pos, ToHeight);
}
```

Once the depth texture is updated with the height information, we can proceed to update the simulation. Set the noise and height texture resource views and set the rain buffer UAV. Call dispatch with: 4, 4, and 1 for the amount of thread groups. Our compute shader defines the noise and height textures in the following way:

```
Texture2D NoiseTex : register( t0 );
Texture2D HeightTex : register( t1 );
```

Our rain buffer UAV is defined as follows:

```
struct RainDrop
{
    float3 Pos;
    float3 Vel;
    float State;
};

RWStructuredBuffer<RainDrop> GridBuffer : register( u0 );
```

We will be using the following constant buffer:

```
cbuffer RainConstants : register( b0 )
{
    float4x4 ToHeight       : packoffset( c0 );
    float3 BoundCenter      : packoffset( c4 );
    float DT             : packoffset( c4.w );
```

```
    float3 BoundSize       : packoffset( c5 );
    float WindVariation    : packoffset( c5.w );
    float2 WindFoce        : packoffset( c6 );
    float VertSpeed        : packoffset( c6.z );
    float HeightMapSize    : packoffset( c6.w );
}
```

Fill the constant buffer with the following values:

▸ ToHeight: Transformation from world space to height map space

▸ BoundCenter: Rain bounding box center position in world space

▸ DT: Time passed since last update

▸ BoundSize: Rain bounding box size

▸ WindVariation: Variation of wind from one drop to the others

▸ WindForce: Wind force value (should change over time)

▸ VertSpeed: Free fall speed

▸ HeightMapSize: Size of the height map texture (128 in this recipe)

Finally, the code for the compute shader is as follows:

```
static const int g_iNumDispatch = 4;
static const int g_iNumThreads = 32;
static const int g_iDimSize = g_iNumDispatch * g_iNumThreads;
[numthreads(g_iNumThreads, g_iNumThreads, 1)]
void SimulateRain(uint3 dispatchThreadId : SV_DispatchThreadID)
{
 // Grap the threads raindrop
 uint GroupIdx = dispatchThreadId.x + dispatchThreadId.y * g_iDimSize;
 RainDrop curDrop = RainData[GroupIdx];

 // Calculate the new position
 curDrop.Pos += curDrop.Vel * DT;

 // Keep the particle inside the bounds
  float2 offsetAmount = (curDrop.Pos.xz - BoundCenter.xz) /
BoundHalfSize.xz;
  curDrop.Pos.xz -= BoundHalfSize.xz * ceil(0.5 * offsetAmount - 0.5);

 // Respawn the particle when it leaves the bound vertically
 if(abs(curDrop.Pos.y - BoundCenter.y) > BoundHalfSize.y)
 {
  // Respawn the particle with random values
  // Sample the noise from the texture
```

```
float4 vNoiseNorm0 = NoiseTex.Load( int3(dispatchThreadId.xy, 0) );
float4 vNoise0 = (vNoiseNorm0 * 2.0) - 1.0;


// Align the position around the bound center
curDrop.Pos.xz = BoundCenter.xz + BoundHalfSize.xz * vNoise0.xy;


// Set the height to a random value close to the top of the bound
curDrop.Pos.y = BoundCenter.y + BoundHalfSize.y;
curDrop.Pos.y -= dot(vNoiseNorm0.zw, 0.2f) * BoundHalfSize.y;


// Set the intitial velocity based on the wind force
curDrop.Vel.xz = lerp(WindFoce, WindFoce * vNoise0.zw,
WindVeriation);
curDrop.Vel.y = VertSpeed;
}


// Check if the particle colided with anything
// First transform the drops world position to the height map space
float4 posInHeight = float4(curDrop.Pos, 1.0);
posInHeight = mul(posInHeight, ToHeight);
posInHeight.xy = 0.5 * (posInHeight.xy + 1.0);
posInHeight.y = 1.0 - posInHeight.y;
posInHeight.xy *= HeightMapSize;


// Sample the height value from the map
float height = HeightTex.Load( uint3(posInHeight.xy, 0) );


// Compare the height of the map value with the drop height
curDrop.State = posInHeight.z < height ? 1.0 : -1.0;


// Write the values back
RainData[GroupIdx] = curDrop;
}
```

Rendering the rain particles requires alpha blending, so it should take place just before the lens flare visibility query. Set the blend state we prepared in the *Getting ready...* section. Set the rain texture to the pixel shaders sampler 0. Set the input layout and vertex buffer to NULL. Set the primitive topology to D3D_PRIMITIVE_TOPOLOGY_TRIANGLELIST and call draw with the amount of rain drops times six.

The constant buffer we use for rendering the rain drops is declared in the shader as follows:

```
cbuffer RainConstants : register( b0 )
{
  float4x4 ViewProj    : packoffset( c0 );
  float3 ViewDir       : packoffset( c4 );
  float Scale       : packoffset( c4.w );
  float4 AmbientColor   : packoffset( c5 );
};
```

Fill the constant buffer with the following values:

- `ViewProj`: Combined view projection matrix
- `ViewDir`: Normalized view direction
- `Scale`: Determines how big the rain drop mesh is going to be
- `AmbientColor`: Ambient color of the rain drops

Our rain buffer is declared in the following way:

```
struct CSBuffer
{
    float3 Pos;
    float3 Vel;
    float State;
};
StructuredBuffer<CSBuffer> RainData : register( t0 );
```

The output from the vertex shader is declared as follows:

```
struct VSRainOut
{
  float4 Pos : SV_Position;
  float Clip : SV_ClipDistance0;
  float2 Tex : TEXCOORD0;
};
```

In order to expend each rain drop into the two triangles, we will use the following constant values:

```
static const float2 arrBasePos[6] =
{
    float2( 1.0, -1.0 ),
    float2( 1.0, 0.0 ),
    float2( -1.0, -1.0 ),
```

```
        float2( -1.0, -1.0 ),
      float2( 1.0, 0.0 ),
      float2( -1.0, 0.0 ),
  };

  static const float2 arrUV[6] =
  {
      float2(1.0, 0.0),
      float2(1.0, 1.0),
      float2(0.0, 0.0),

      float2(0.0, 0.0),
      float2(1.0, 1.0),
      float2(0.0, 1.0),
  };
```

Our vertex shader uses the following code:

```
  VSRainOut VS_Rain(uint VertexID : SV_VERTEXID)
  {
      VSRainOut output;

      // Get the base position
      float3 pos = RainData[VertexID / 6].Pos;

      // Find the expension directions
      float3 rainDir = normalize(RainData[VertexID / 6].Vel);
      float3 rainRight = normalize(cross(ViewDir, rainDir));

      // Extend the drop position to the streak corners
      float2 offsets = arrBasePos[VertexID % 6];
      pos += rainRight * offsets.x * Scale * 0.025;
      pos += rainDir * offsets.y * Scale;

      // Transform each corner to projected space
      output.Pos = mul(float4(pos, 1.0), ViewProj);

      // Just copy the UV coordinates
      output.Tex = arrUV[VertexID % 6];

      // Clip particles that collided with the ground
      output.Clip = RainData[VertexID / 6].State;

      return output;
  }
```

Our pixel shader will sample from the rain drop texture through the following texture and sampler declarations:

```
Texture2D RainStreakTex : register( t0 );
SamplerState LinearSampler : register( s0 );
```

Finally, the following code is used by the pixel shader:

```
float4 PS_Rain(float2 tex : TEXCOORD0) : SV_Target
{
    float fTexAlpha = RainStreakTex.Sample(LinearSampler, Tex).r;
    return float4(AmbientColor.rgb, AmbientColor.a * fTexAlpha);
}
```

How it works...

Rain simulation is considered easy because the drops are pretty uniformly distributed and they fall in a very specific direction. As we only care about rendering stuff that is visible to the camera, rain can be rendered and simulated only inside a volume in front of the camera. Each time the camera moves, the volume moves with it and so do the rain drops. It is important to avoid a tight volume around the camera otherwise any camera movement can cause the rain simulation to look wrong. Another important aspect is the noise texture which has to contain high frequency noise that covers all 256 values evenly. If the noise texture is not even, you will notice gaps where drops never show up.

Simulating the rain is handled in two separate steps. First, we prepare the height map for collision detection between the rain drops and the scene, then we perform the actual simulation update. Using a height map for collision detection prevents rain from falling through roofs or other scene elements that should otherwise block the rain. As you may have noticed from the shader we used for rendering the scene into the height map, the vertex shader is similar to the one we used for shadow map generation. The only thing new about this step is that we need a different matrix. As usual, this matrix is a combination of a view and projection matrices. The view matrix is a transformation that points straight down from the top of the rain bounds into its center as shown in the following illustration:

$$\begin{bmatrix} 1 & 0 & 0 & 0 \\ 0 & 0 & -1 & 0 \\ 0 & 1 & 0 & 0 \\ -P_x & -P_z & -P_y & 1 \end{bmatrix}$$

The projection we use for the height map is an orthographic projection that wraps around the rain bounds. As a reminder, orthographic projection matrix has the following form:

$$
\begin{bmatrix}
2/w & 0 & 0 & 0 \\
0 & 2/h & 0 & 0 \\
0 & 0 & 1/\ (zf\text{-}zn) & 0 \\
0 & 0 & zn/\ (zn\text{-}zf) & 1
\end{bmatrix}
$$

Use the bounds X value for width, Z value for height, and Y value for Z far (**zf**). Z near (**zn**) is zero. Once all the scene meshes that intersect with the rain bounds are rendered, the depth texture contains the non-linear height values in the rain space.

Before compute shaders and UAVs where introduced, there was no way to read and write to the same memory on the GPU. A common way around this problem was double buffer the data using two textures, so the previous update values were read from one texture and the current update results were written to the second texture. UAVs support both read and write operations to the same memory, so our simulation can use a single buffer. Our update code is split into three steps: update the rain drop position, re-spawn the drop if it's outside the vertical bound, and collision test with the height map.

Updating the position is based on explicit Euler integration. For each drop, the new position is equal to velocity multiplied by the time that passed from the last frame plus the previous position (the velocity is fixed from the moment the rain drop is spawned). Once the new position is set, we make sure all the rain drops stay inside the bounds by

Once the position is fully updated, we check if the drop is outside the vertical bounds and re-spawn it if it is. The noise map is used for generating semi random values when spawning a new rain drop. When generating the velocity values, wind force contribution is taken into account to create variation in each rain drop trajectory. Wind force values should change between frames to create variations in the rain direction over time.

Collision detection is handled in a similar way to shadow mapping. The world position of the rain drop is transformed into the height map space. We sample the depth straight from a pixel coordinate without a sampler, so the rain drop position in the height that maps clip space had to be normalized and multiplied by the pixel size of the height map. If the sampled depth is higher than the rain drop depth, it means the drop is above the scene and there is no collision.

After the rain buffer is updated, we proceed with the usual rendering up until the lens flare rendering where we render the rain drops. The reason for this location is that we want the rain drops to render after all the scene elements are added to the depth buffer and the scene is fully lit, but before the lens flare so that the flares render on top of the rain drops.

There are two techniques we use for rendering the rain drops that we didn't use up to this recipe: clip planes and directional extrusion. Each rain drop contains a state value which indicates if it collided with the scene. Drops that did collide with the scene get removed right after the vertex shader is executed using `SV_ClipDistance0` which takes a single float value: negative will cull the vertex and positive will render it as usual. We pass in the drops state value which will clip rain drops that collided with the scene.

Converting each rain drop into the two triangles is done based on the particle's position and velocity. We use the cross product to find a normalized vector perpendicular to the velocity and camera direction. Along with the normalized velocity direction, we use these two vectors to create two triangles which face the camera and have a rectangular shape longer in the velocity direction.

One thing to keep in mind is that even though we simulate all the particles all the time, you can render only a portion of them, so you can control the rain amount anywhere between light and heavy rain fall.

Index

S

screen space ambient occlusion **143-153**
screen space effects
  ambient occlusion 143-153
  lens flare 154-160
  reflections 160-167
  sun rays 168-175
shader resource view (SRV) **144, 169**
shadow caster **72**
shadow-casting support
  adding, to point light 85-91
  adding, to spot lights 73-84
shadow mapping **73**
shadow maps
  visualizing 105-107
shadow receiver **72**
SOSetTargets function **182**
specular light **16**
SpotCosInnerCone **21**
SpotCosOuterCone **21**
spot light
  about 20, 21
  shadow-casting support, adding to 73-84
  texture, projecting 29-32
  working 22

spot light volume
  about 65
  generating 65-69
SRV (shader resource view) **111**
Stencil Shadows **72**
sun rays **168**

T

tone mapping technique **122**

U

UAV (Unordered Access View) **53, 111, 144**

V

varying penumbra size, PCF **99-104**

X

Xbox 360 **40, 102**

About Packt Publishing

Packt, pronounced 'packed', published its first book "*Mastering phpMyAdmin for Effective MySQL Management*" in April 2004 and subsequently continued to specialize in publishing highly focused books on specific technologies and solutions.

Our books and publications share the experiences of your fellow IT professionals in adapting and customizing today's systems, applications, and frameworks. Our solution based books give you the knowledge and power to customize the software and technologies you're using to get the job done. Packt books are more specific and less general than the IT books you have seen in the past. Our unique business model allows us to bring you more focused information, giving you more of what you need to know, and less of what you don't.

Packt is a modern, yet unique publishing company, which focuses on producing quality, cutting-edge books for communities of developers, administrators, and newbies alike. For more information, please visit our website: www.packtpub.com.

Writing for Packt

We welcome all inquiries from people who are interested in authoring. Book proposals should be sent to author@packtpub.com. If your book idea is still at an early stage and you would like to discuss it first before writing a formal book proposal, contact us; one of our commissioning editors will get in touch with you.

We're not just looking for published authors; if you have strong technical skills but no writing experience, our experienced editors can help you develop a writing career, or simply get some additional reward for your expertise.

UnrealScript Game Programming Cookbook

ISBN: 978-1-84969-556-5   Paperback: 272 pages

Discover how you can augment your game development with the power of UnrealScript

1. Create a truly unique experience within UDK using a series of powerful recipes to augment your content

2. Discover how you can utilize the advanced functionality offered by the Unreal Engine with UnrealScript

3. Learn how to harness the built-in AI in UDK to its full potential

XNA 4.0 Game Development by Example: Beginner's Guide

ISBN: 978-1-84969-066-9   Paperback: 428 pages

Create exciting games with Microsoft XNA 4.0

1. Dive headfirst into game creation with XNA

2. Four different styles of games comprising a puzzler, a space shooter, a multi-axis shoot 'em up, and a jump-and-run platformer

3. Games that gradually increase in complexity to cover a wide variety of game development techniques

Please check **www.PacktPub.com** for information on our titles

www.ingramcontent.com/pod-product-compliance
Lightning Source LLC
LaVergne TN
LVHW062314060326
832902LV00013B/2213